Paramagnetism

Rediscovering Nature's Secret Force of Growth

Paramagnetism

Rediscovering Nature's Secret Force of Growth

Philip S. Callahan, Ph.D.

Acres U.S.A.
Greeley, CO

Paramagnetism

Rediscovering Nature's Secret Force of Growth

Acres U.S.A.
P.O. Box 1690
Greeley, CO 80632 U.S.A.
(970) 392-4464
info@acresusa.com • www.acresusa.com

Printed in the United States of America

Callahan, Philip S., 1923-
    Paramagnetism: rediscovering nature's secret force of growth /
    Philip S. Callahan. — Greeley, CO, ACRES U.S.A., 1995.
    xii, 132 p., 23 cm.
    Includes index.
    ISBN-13: 978-0-911311-49-5

    1. Growth (Plants) 2. Agriculture. 3. Paramagnetism.
    I. Callahan, Philip S., 1923- II. Title.

QK745.C35 1995   581.3
                                    QBI95-20336

Library of Congress Catalog Card Number: 95-078655

By the Author

For Winnie

Rocks are known for their stability. Christians well know that Christ said, *Upon this rock I shall build my Church*. We may know then that love is like a rock — a stable phenomenon that turns into a "soil" for the nourishment of mankind. This is my most important book, and my rock of nourishment is my greatest earthly love, and so I dedicate this book to her.

Cape May Warbler
and Little Blue Butterfly

The Cape May Warbler is misnamed as it is seen far more often during the spring and fall migration in Florida and Georgia than at Cape May, New Jersey. The Cape May breeds in the far north of Canada, nesting in the upper branches of fir and spruce trees. It migrates across the United States, usually arriving in Key West by November. Although it is rarely seen except during migration, when it funnels across the eastern United States and into Florida, I did observe one with a huge flock of other warbler species at Little Switzerland, North

Carolina on August 19, 1975, indicating that these Canadian breeding birds leave the far north early in August. The beautiful spring colored male is most often seen around town and village lawns in Florida during April and May, the time of the spring migration north.

This bird obtained the unsuitable name of Cape May Warbler from Alexander Wilson. He named it from a specimen collected in 1811 at a maple swamp near Cape May, New Jersey by his friend George Ord. Wilson was, in truth, the greatest American ornithologist. He is less well known than John J. Audubon, because Audubon was also a great painter and a favorite of bird artists.

The Cape May winters in the Bahamas and West Indies. It may be headed for extinction thanks to insecticides spread about lands below our southern borders by American corporations. There has been a 50% drop in warbler counts over the past 20 years. Healthy soil could stop this.

The Little Blue Lycaena butterfly is common across the entire United States. *Lycaena pseudargiolus* has many subspecies, each exhibiting color variations. In general, it is pale blue with black wing margins. It feeds on the petals of dogwood blossoms.

Contents

Author's Note

For the scientifically and mathematically inclined, the best scientific treatise on paramagnetism is *Paramagnetic Relaxation*, by Dr. C.J. Gorter of the Zeeman Laboratories of the University of Leyden in the Netherlands.

Dr. Gorter, a professor of experimental physics at the University of Leyden, was in his thirties when the book was written between 1944-45. During that time he was unable to experiment, so the treatise is highly theoretical and mathematical. He was "confined" to one room of the Zeeman Laboratories during these years, known as the starvation years in Holland. Not only had Hitler depleted the food supplies of his victims, but males between 20 and 40 were regularly picked up in the streets and sent to labor camps.

Dr. Gorter was forced to hide out in the Zeeman Labs and rarely ventured outside. The laboratory was safer than his home, which could be invaded at any moment. He wrote the entire work while so confined. Fortunately he survived the war. I think of him as a lucky, scientific, Anne Frank.

It was his chapter on "Crystalline Fields" and the interaction between magnetic ions that led me to look at the paramagnetic force in rock and then on to soil, which is eroded rock. After all, rocks are crystalline.

I don't know if Dr. Gorter is alive today, but alive or dead, I dedicate this work to his memory. It is fitting that a force that in the long run (if farmers will only listen) could prevent starvation all over the world, was first delineated by a victim of starvation. It is also ironic that Dr. Gorter himself wrote, "The study of paramagnetic relaxation is not an important chapter in modern physics." It still is not!

Acknowledgement

A very special thanks to Dr. Robert Wilkinson, entomologist, and Dr. Jeffrey Klein, M.D. for their constant help in the fields and woods of Florida, digging soil and measuring frequencies, and to Kenneth Silver who had the fortitude to follow me into the jungles of the Amazon in order to measure river banks and tree canopy frequencies. Also a special thanks to Ann and Charles Walters and Fred Walters for years of encouragement and helping publish my life's work.

Lastly, without the help and encouragement of many friends, this book would not have been written. Most important are Lee Leitner, Dr. Edward O'Brien, Robert Pike, Dr. Arden Andersen, and Harry Kornberg.

Foreword

The fundamental trait that separates Phil Callahan from most other members of his profession is that of conceit — Callahan's total lack of it. Most scientists believe their extreme specialization is the high road to enlightenment. They believe they are above nature and have the right to manipulate, to re-engineer nature. And they believe they alone are the keepers of the keys of knowledge for their narrow fields of focus. Callahan is different.

His every move reflects a cross-pollination of ideas. Always open to new thoughts, whether from a scientist or a farmer, from nature or from the ancients, he makes sense of seemingly incongruous areas of knowledge. Quite often the cross-pollination process begins in his own mind, drawing from his own rich, diverse background. A painter and a mathematician, a writer and a physicist, a poet and a biologist. When it comes to left-brain and right-brain thinking, he might well be the first man with two brains.

Ever humble, he adores nature and God's beautiful systems. He respects those peoples who have come before us, seeking to understand them, not to judge. And he truly loves farmers, their vastly underappreciated role in our society notwithstanding.

For Phil Callahan, scientific discovery starts first with an observation of nature. Then his computer-like brain locates the common thread that links that observation with others. More often than not, no one else has found that thread. It is through the magical combination of his background in physics, biology, chemistry, geology, archaeology, astronomy, history and religion that the backbone of this book — *Paramagnetism* — came about.

The author believes with all of his heart that this is the most important book of his life. It is the book that can save agriculture around the world.

He kept it short, so as not to overwhelm. He kept it direct, to be easily understood. It is the culmination of a lifetime of exploration. This book identifies a force long known of in physics and its incredible power and importance to the greatest endeavor on earth, farming.

But in the end, it is up to the reader to study, understand, experiment and implement these findings. Beyond the healthy growth that will most certainly arise out of this book, another benefit could appear. A little bit of Phil Callahan's love of life just might rub off as well.

— Fred C. Walters

I paid a dime for a package of seeds.
The clerk tossed them down with a flip.
"We have them assorted to every man's needs,"
He said, with a smile on his lips —
Pansies, poppies, asters, and peas —
Ten cents a package; now pick as you please.

Now seeds are just dimes to the man in the store,
And dimes are the things that he needs,
And I've been to buy them in the store before,
And thought of them merely as seeds.
But it flashed through my mind as I took them this time:
You've purchased a miracle here for a dime!

You've a dime's worth of power which no man can create;
You've a dime's worth of life in your hand.
You've a dime's worth of mystery . . . destiny . . . fate,
Which the wisest cannot understand.
In this bright little package — now isn't it odd?
You've a dime's worth of something known only to God!

— *Anonymous*

PART I
SOIL

The Soil and Health

The usual subdivision of science into chemical, physical, botanical, and other departments, necessary for the sake of clarity and convenience in teaching, soon began to dominate the outlook and work of these institutions. The problems of agriculture — a vast biological complex — began to be subdivided much the same way as the teaching of science. Here it was not justified, for the subject dealt with could never be divided, it being beyond the capacity of the plant or animal to sustain its life processes in separate phases. It eats, drinks, breathes, sleeps, digests, moves, sickens, suffers or recovers, and reacts to all its surroundings, friends and enemies in the course of twenty-four hours. Neither can any of its operations be carried on apart from all the others; in fact, agriculture deals with organized entities, and agriculture research is bound to recognize this truth at the starting point of its investigations.

In doing this, but adopting the artificial divisions of science as at present, established conventional research on a subject like agriculture was bound to involve itself and magnificently has got itself bogged. An immense amount of work is being done, each tiny portion is a separate compartment; a whole army of investigators has been recruited; a regular profession has been invented. The absurdity of team work has been devised as a remedy for the frag-

mentation which need never have occurred. This is nonsensical. Agricultural investigation is so difficult that it will always demand a very special combination of qualities which from the nature of the case is rare. A real investigator for such a subject can never be created by the mere accumulation of the second rate.

Nevertheless, the administration claims that agricultural research is now organized, having substituted that dreary precept for the soul-shaking principle of that essential freedom needed by the seeker after truth. The natural universe, which is one, has been halved, quartered, fractionized, and woe betide the investigator who looks at any segment other than his own! Departmentalism is recognized in its worst and last form when councils and supercommittees are established — these are the latest exercises — whose purpose is to prevent so-called overlapping, strictly to hold each man to his allotted narrow path and above all to enable the bureaucrat to dodge his responsibilities.

The Soil and Health, *1947*
Sir Albert Howard

4　*Paramagnetism*

Introduction
BREAD FROM STONES

Years ago, while I was enjoying leisure travels in Ireland, I picked up a book titled *Farming and Gardening for Health and Disease*, by Sir Albert Howard. It was read, or I should say scanned, so quickly that I retained only dim memories of its content. What I do remember is that it dwelled on farming techniques in India; techniques like those which have been in use by the Irish since ancient Celtic times.

My bible for pre-World War II Irish farming is the masterpiece *Irish Heritage*, by the astute professor of Queens University, Belfast, E. Estyne Evans. It is the only book I have two copies of — one in my lab, and one in my den.

I had, between 1944 and 1946, lived in Fermanagh County, located in Northern Ireland on the Denegal border. I was, in truth, a sergeant in the Army Air Corps, but may as well have been a 19th-century potato farmer, for I was far more interested in the natural history and agriculture of the beautiful Erne Valley than I was in the exactitudes of my job as a low-frequency radio-range technician. These were the years before the intrusion of deadly chemical farming.

Ideas that are meaningful develop slowly. Sometime, long after I left Ireland, I came to realize that plants, insects, and soil and the nebulous photons of electronic systems were all in one. Everything is connected to everything else, especially by the electromagnetic spectrum.

Even in my younger days, if I had really paid attention to what Sir Albert was writing about, I would have realized that his life's work was built on a solid foundation and keen understanding of the inner conectedness of all of nature. He had little use for the reductionist methods of modern science. More than any other agricultural scientist of his time, he understood that reductionism, like communism, might well lead to the destruction of viable agriculture.

Physics is the science that connects chemistry to biology. That being so, a scientist that does not have a basic understanding of physics is more ignorant of life than a leaping flea hopper, which at least knows it must jump (physics) to feed (chemistry) on a plant (biology).

In the mid-1970s, I was delighted to learn that my editor, Devin Garrity of the Devin-Adair Company, had published Sir Albert's book in the United States. He gave me a copy of the book under its new title: *The Soil and Health*. Needless to say, I studied it more thoroughly this time around. *The Soil and Health* is about the biological/chemical makeup of agriculture. It is one of the original treatises on composting and crop rotation. It also talks about the mess that the German chemist Justus von Liebig started in agriculture with his "pure" chemistry concept of plant growth.

Justus von Liebig, a laboratory chemist, equated life with N, P & K (nitrogen, phosphorus and potassium). I doubt very much if, as a lad, he ever chewed on a fresh piece of grass while cloud watching on his back. He published *Chemistry in the Application to Agriculture* around 1840. His book preceded the great book on soil formation by Charles Darwin titled *The Formation of Vegetable Mould Through the Actions of Worms, with Observations of Their Habits*. This book shines with Darwin's aura of genius, unlike his treatise on the unprovable theory of evolution. God probably has a hundred different ways of creating life, evolution being only one among many. Like creationism, it is a reductionist either/or science. Because Darwin, like von Liebig and other biological chemists, overlooked the elegant work of the "two T's" — John Tyndell and Nikola Tesla — he failed miserably in his understanding of natural forces. It remained for another chemist, Julius Hensel, to point the way with his beautifully titled book, *Bread from Stones*. He also had little use for the concepts of von Liebig. He talks about von Liebig's mistake:

> *Very simply, von Liebig was the first agricultural chemist. He found that the ashes which remained from grain mainly consisted of phosphate of potassa. From this he concluded that phosphate of potassa must be restored to the soil, and that was very one-sided. Von Liebig had forgotten to take the straw into account, in which only small quantities of phosphoric acid are found, because this substance, during the process of maturing, passes from the stalk into the grain. If he had not only calculated the seed but also the roots and the stalks, he would have found what we know at this day, that in the whole plants there is much more lime and magnesia as potassa and soda, and that phosphoric acid forms only the tenth part of the sum of these basic constituents. Unfortunately von Liebig also was of the opinion that potassa and phosphoric acid have to be restored to the soil as such, while anyone might have concluded that instead of the exhausted soil we must supply earthly matter from which nothing has been grown. Such untouched*

earthly material of primitive strength we get by pulverizing rocks into which potassa, soda, lime, magnesia, manganese and iron are combined with silica, alumina, phosphoric acid, fluorine and sulfur. Among these substances fluorine, which is found in all mica-minerals, has been neglected by von Liebig, by all his followers, and has never been contained in any artificial manure. But as we know from late investigations that fluorine is regularly found even in white and yellow birds' eggs, we must acknowledge it is something essential to the organism. Chickens get this fluorine and other earthy constituents when they have a chance to pick up little slivers of granite. Where this is denied them, as in a wooden hen house, they succumb to chicken cholera and chicken diphtheria.

The key to *Bread from Stones* is contained in this one paragraph where he says, "We must supply earthly matter" and later "little slivers of granite." His use of the word granite implies that he not only knew that good soil is made from eroded stone, but which kind of stone is best suited to a viable agriculture. Because he was a chemist, like von Liebig, he emphasized the chemical constituents of stone, and therein lies the crux of what this and subsequent chapters are about.

It is a dismaying fact that there are some rocks, including some granite, that when ground up and added to the soil accomplish little for plant growth even though they may contain all the aforementioned chemicals. The simple and irrefutable fact is that if the force called *paramagnetism* is not present, little benefit will accrue to the soil in spite of the fact that certain proportions of the chemicals are present. This book is about that force, one among many natural forces, but the one most often missing from poor soil around the world.

Poor soil produces sick plants. Since insects are scavengers, they are attracted to both old and sick plants. They are nature's recyclers. Crop plants should be young as they are harvested to eat. If they are young and healthy, insect losses are almost nil. If they are young and sick, losses from insects can be devastating.

We will deal then with growing healthy plants, but it does not imply excluding the need to learn about the other two necessities of agriculture: composting and soil organisms. Books on composting and soil organisms abound. Add them to these words and all of the ingredients of a flourishing and viable agriculture will attain. *C*omposting, *O*rganisms of the soil, and the *P*aramagnetic force (COP) might well prevent a worldwide famine from destroying mankind.

One last thought for farmers, but it equally applies to all of an ecological mindset, especially those in positions of leadership who are attempting to save our wilderness and our wildlife from destruction. Because plants are healthy and do not attract hoards of insects does not mean that insectivorous wild creatures like birds will suffer. The hoards of red-winged blackbirds that descend on corn infested with corn earworm larvae not only further damage the corn, but would live more healthfully in the cattail marshes of their habitat, free from crops contaminated by insecticides.

One half of the population of North American wood warblers has disappeared over the last twenty years. This is probably due, in part, to insecticides sold freely by the U.S. agricultural chemical industry to the Central and South American countries to which North American warblers migrate. With few government controls, in certain southern regions insecticides are spread by the tons. There are now more pounds of pesticides per acre utilized worldwide than when Rachel Carson wrote *Silent Spring*, that epic treatise on the subject.

Insecticides and weed killers are the modern curse of environmental health. We never did need them. The simple fact is that they both destroy viable soil. A healthy soil ecology through healthy agriculture means a healthy world population reinforced by that most democratic foundation — the family farm.

10   *Paramagnetism*

Chapter 1
VOLCANOES

Sometimes rocks speak quite sharply. Whenever there's sufficient heat, pressure, and water to melt great masses of rocks, they may be expected to intrude the sediments above them. Quite frequently this hot material reaches the surface with explosive results as expanding steam produces violent volcanic eruptions.

Dance of the Continents
John W. Harrington

Science believes that agriculture began thousands of years before Christ somewhere in the Middle East. This may well be true of man's part in agriculture, but God's part began when he created the volcano.

I well remember the newspaper accounts of the eruption of Mount St. Helens in Washington. It is a twin cone to beautiful Mount Hood in Oregon. Earthquakes began on March 20, 1980, but the real boom did not occur until May 18. David Johnston, the volcano expert with the U.S. Geological Survey, radioed his headquarters a short message: "Vancouver! Vancouver! This is it." The volcano lover Harry Truman, a mountain man of the region, refused to take warning and perished in an old mine tunnel. David Johnston also perished.

Mount St. Helens erupted with a force calculated to equal fifteen hundred Hiroshima-size atomic bombs. Over one and a half cubic miles of rock were blown across the Washington countryside, lowering the peak of Mount St. Helens from a height of 9,677 feet to a height of 8,377 feet — 1,300 feet of the cone disappeared! The blast was heard 200 miles away, and volcanic ash was blown downwind in a broad belt that spanned an area from Mississippi in the South to Canada in the North.

My own experiences with volcanoes involve three climbs of Mt. Fuji and many days in Pompeii dreaming about what it must have been like to survive such an explosion. Pliny the Elder, another volcanic observer, died in that eruption.

In Japan, my sister Ann and I came close to perishing in the eruption of Mount Asama near Karuizawa in the Japanese Alps. That mountain, so we were told on the day of our projected climb, last erupted 500 years ago and was highly unlikely to erupt on the day of our planned climb. It did! Fortunately, I had been called back to Tokyo just prior to the eruption. That mountain spewed a column of ash thousands of feet high across the Japanese Alps. The explosion was heard over a hundred miles away. My sister was left alive to have three beautiful daughters and I to write a book on paramagnetism. That was in 1947, a time when I little understood that volcanoes are not disasters but blessings in disguise. No volcanoes, no agriculture — for volcanic ash and rock are the guts of good soil.

What I best remember in the newspaper reports of Mount St. Helens was the great concern about the loss of crop land and forests buried in ash, and the fact the pigs and snakes seemed to know it was about to occur. To this day no one has decoded the earthquake/volcano mechanism of pig and snake prophecy, but in this book we can learn at least one of the great benefits of volcanic explosions and about the magic ash that it puts down across the countryside.

A few years after the eruption of Mount St. Helens, articles began to appear, most written in adjectives of great surprise, detailing how fast the forests were returning, the plants popping up, the streams revitalized and even nearby farmers delighted with their crop output. Apparently modern man and agribusiness had forgotten that good soil comes from volcanic rock and not the chemical industry. Mount St. Helens demonstrated that God knows what He is doing and corporate America only believes it does. It is not that I believe corporate agriculture is evil, only misguided. Perhaps they can learn from God's volcanoes.

E.A. Vincent, in the marvelous *Forces of Nature*, makes only one statement that I can take exception to. It is in the first sentence of his masterful dissertation on volcanoes. He states: "Since in close quarters a volcanic eruption is an impressive energetic phenomenon, but being narrowly localized in space and time, it is rather superficial on the scale of the whole earth." He further elucidates: "As my professor of geology once remarked to the class of first-year students of which I was a member, 'Would you conclude from the presence of a boil on the back of your neck that your whole body was filled with pus?'"

In these quotes we can see the rather peculiar reasoning of a specialist. No medical person that I know believes that a boil explodes and spreads a fine dusty mist of matter, good or bad, across the entire body — yet this is precisely what a volcanic explosion of the "boil" on the earth's surface does. It does indeed affect thousands upon thousands of miles of the earth's surface. The volcano is a powerful exploding, soil-dispersing "boil."

In the next paragraph Dr. Vincent redeems himself as he states in no uncertain terms: "Man has always feared volcanic manifestations as forces beyond his control, bringing death and destruction. But he has also gradually learned that they may bring some compensating benefits: rich volcanic soils; deposits of minerals and ores; and more recently, potential sources of unstable heat and energy. The very existence of volcanoes, as of the earthquakes almost invariably associated with them, indicates that the planet Earth is not a static body but dynamic organism in a constant state of change and evolution." Note that he places the words "rich volcanic soil" first, even before the first love of industry — deposits of minerals and ores!

The physics of volcano formation is easily summarized. Specifically, a volcano is a surface phenomenon that follows large earthquakes triggered by a subsurface movement of what geologists call tectonic plates. The earth is made up of a number of such huge plate-like platforms floating on a sea of viscous, flowing, iron-rich, glassy rock called magma. The more brittle crust rock on top moves about separating, colliding and triggering powerful interactions that, over eons of time, effect the physical makeup of the earth's surface. These moving continental plates are the mountain and ocean builders of time. Volcanic cones usually arise and form in chains where the denser subsurface mantle rock is carried by a portion of the molten magma. Theorists believe that the thermal energy required for convective movement within the mantle, and the melting of the viscous material, is caused by the disintegration of small amounts of the radioactive elements of uranium, thorium and potassium contained within the earth surfaces. In other words, complex and terrifying energies of molten rock generate steam that expands, and under tremendous pressure, bursts through the surface crust, forming volcanic cones and spewing out, so to speak, our most fertile soils.

Few realize that if the theorists are correct, our earth is one huge soil-forming, atomic-steam powerplant. In that case, a volcano is, in reality, the steam explosion of an atomic energy system. The silicate formed by the partial atomic-steam explosion and thrown off by active volcanoes, is cooled in the atmosphere to form a rock called basalt. A moving plate curves down beneath a crust plate where friction, added to the atomic pump, feeds the melt. This energy cracks the earth under pressure and carries the liquid basalt rock upward where dust and rock forms the cone and the folds around the cone.

Of course, this is a simplified description of the whole process that is complex beyond all imagination. It has, however, been researched by numerous geologists over the years since the German meteorologist, Alfred Wegener, in a book titled *The Origins of Continents and Oceans*, came up with what is now known as the tectonic theory of land formation. He was, of course, ridiculed in his own day.

16 *Paramagnetism*

Chapter 2
FROM ROCK TO SOIL

German scientist, Ewald Wollny (1888) has been called the "pioneer of soil and water conservation research" (Baver, 1938). In the last quarter of the 19th century, he made extensive investigations of the physical properties of soil that effect runoff and erosion. He studied the effect of various factors, including steepness of slope, plant cover, soil type, and direction of exposure, on runoff and erosion from small plots under natural rainfall. He also studied factors affecting percolation, transpiration, and evaporation from soils, and he investigated effects of compaction on the physical properties of soil. However, Wollny's discoveries were apparently overlooked by American researchers until the mid-1930s. (Nelson, 1958)

Estimated Erosion and Sediment Yield
USDA Proceedings, L.D. Meyer

There is nothing new about the original work of an innovative and non-reductionist scientist being overlooked, whether German, English, American, Hebrew, or Arab. In fact, most innovative Arab researchers of past centuries have not only been overlooked but completely discarded by the West.

The only thing the Crusaders ever accomplished, other than killing a lot of innocent people, was to bring Arab science to the West. In the 12th century, the crusader Frederick II invented bird banding in order to investigate bird migration. He was also the first to use nicotine sulphate to kill bird lice on the wings of his hunting falcons. I have never seen credit given to him for either of these modern scientific endeavors. The Arab people understood bird migration and poisons of all sorts long before western Europeans ever thought about such fields of study. Frederick II learned much from his contacts with Middle Eastern Arabs. He went east to fight, and after discovering that killing in the name of God is a dastardly occupation, he remained to study. He became a good friend of the sultan he was supposed to kill — sometimes things do turn out right in life. The white flowing robes and grey beards of Arab and Hebrew philosophers from Jerusalem, Baghdad and Syria were a common sight at Frederick's court at Apulia in southern Italy.

Frederick's favorite castle is still in existence. The Castle del Monte sits on the flat, desert-like plane of Tavoliere in the region of Apulia not far from the town of Foggia. The flat, rock covered, dark earth region of desolation stretches away in all directions from his hilltop castle. Modern man is prone to wonder why Frederick favored such a non-agricultural region. It is green only during a very short winter rainy season. He favored it because it was a huge open region for hunting his falcons. He was, however, quite as interested in agriculture as in falconry.

His love of farming is attested to by the fact that he kept vast herds of sheep, cows, goats, and pigs, as well as bees and pigeons. On his more soil-rich estate, he grew oats, millet, hemp, cotton, corn, wheat, grapes and olives. His advisors were the Cistercian monks who were also the scientific farmers of those days. These monks developed new breeds of animals and plants using crossing methods later perfected by the monk Mendel and now called the science of genetics. The

info@acresusa.com
970-392-4464
https://www.acresusa.com

1-12294
Order Date: Aug 21, 2019

SHIP TO: Thomas Swafford
856 Bolton Abbey Lane, Vandalia, OH 45377
P: +17606834998 **E:** delta_wye_power@yahoo.com

Requested Shipping: Economy Mail (7-10 days)

SKU	Desc	Price	Qty
6158	Paramagnetism	$10.50	1
6103	The Enlivened Rock Powders	$10.50	1

Subtotal:	$21.00
Discount:	$0.00
Shipping:	$3.95
Tax:	$0.00
TOTAL:	$24.95

Thanks for choosing Acres U.S.A. The future of food is non-toxic!

Cistercians experimented with soil types as did the father of the study of soil erosion, Ewald Wollny.

Leaving those ancient times, when the monks worked with nature instead of against it, and leaving as well, the elegant work of Wollny between 1877 and 1895, most of the significant work on erosion has been accomplished in the United States by the U.S. Department of Agriculture (USDA). It was not until 1907 that the USDA declared an official policy covering land protection. This is an important date in the evolution of science. Soil is ecology!

Erosion carries soil away from its original location. As we have seen, erosion begins with volcanoes. This may lead one to ask, if erosion is so bad, and volcanoes are erosion producers, why praise volcanoes? Is erosion good or is erosion bad? That is like asking, are weeds good or are weeds bad? It depends on where the erosion ends up and where the weeds are located. This may be true in the short-term, but in the framework of centuries, weeds and erosion are most certainly good, for without them we could not have life on earth as we know it today.

Most Chinese soil came from the dry hills of inner China, blown eastward over millions of years. Such windblown dirt is called loess soil. China soil was also badly eroded down the mountain slopes to rivers where it was swept along to build banks or fill the huge fertile estuaries of the far off rivers. The soil that was laid down across the flat land of China hundreds upon hundreds of thousands of years ago has, for 4,000 or more years, fed a population that has risen from 400 million after World War II to more than one billion today — China has a big problem!

In America today, soil laid down from Rocky Mountain volcanic action and deposited many, many feet thick, is sometimes blown away by wind erosion, as it was in the Dust Bowl regions during the 1930s. Americans also have a big problem.

It is not based on excessive human population, but rather on present-day wind erosion.

Soil erosion by wind occurs in areas where yearly rainfall is minimum. The great, dry central American prairie measures about twelve inches of rain per year. Wind erosion is an ongoing problem where green top cover is disturbed by agricultural processes. Large movements of soil occur not only where rainfall is scarce, but also where there are strong prevailing winds at all levels, from high altitudes down to ground level. These regions are associated with massive flatland areas.

Water erosion is most often associated with high rainfall, and hilly or mountainous regions drained by creeks and rivers.

In the United States, soil scientists have plotted a map called the rainfall erosion index for erosivity. They use it to predict erosion on a yearly basis over the country. It is easily observable that the index decreases from a high value of 450 to 550 along the Gulf Coast, where there is a high average rainfall, to a low of 20 in the high, dry plateau of Nevada and

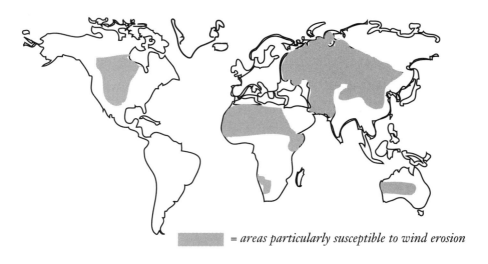

= *areas particularly susceptible to wind erosion*

The flatlands of America are composed mainly of fine soil from sedimentary rock that is easily blown by prevailing winds. Lack of hedgerows or trees allows for high wind erosion.

Utah. From such maps erosivity — and thus loss of soil by rain — can be determined from any location in the United States. In the eastern United States, we observe that most erosion is caused by rainfall and not wind.

Erosion is a continuing process. It begins with soil laid down by volcanic action and continues by a process called *weathering*. As important as weathering is to soil formation and dispersion, or erosion, I could not find the word in either the index of *Soil*, the 1957 *Yearbook of Agriculture*, or in any of the dozen books I have on soil, rocks and minerals. Weathering is the method by which solid rocks are turned into the smaller particles that make up our soil. Classifications of soil particle shape and size are complex but have been well studied. Formation of soil begins with two different types of weathering: physical and chemical.

Rocks are divided into three different types according to their geological origin. *Igneous* rock is formed by the cooling

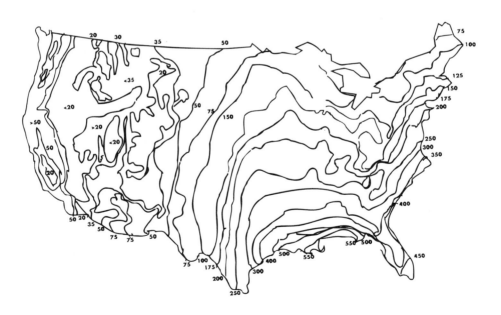

Map of average annual values of rainfall erosion index. Wischmeier and Smith (1978).

and hardening of the viscous magma. It is classified according to its decreasing silica content, and acid to basic.

Rocks formed under conditions of temperature, pressure and chemical reaction that do not involve melting, as in the case of magma, are called *metamorphic* rock. Granite may be igneous or metamorphic. In general, metamorphic rock is formed deep within the earth's crust. It is associated with the great pressures and temperatures involved in mountain building. Metamorphic means a change in form. Recrystallization is the term for a changed particle size and structure during the process of metamorphism. Gneiss, schist and slate are examples of metamorphic rock.

The third major classification of rock is known as *sedimentary* rock. All rock disintegrates over the eons as a result of mechanical forces (water, ice, wind and friction) and chemical weathering. These rock particles, in the form of clay, silt, sand, gravel, and their dissolved minerals, are transported by erosion, by wind and water, to a new location where they are deposited in layers. The particles may eventually be cemented together forming what is called *clastic* sedimentary rocks. The manner in which chemical sedimentary rock is formed and deposited allows geologists to reconstruct the earth's geological events.

Igneous minerals, like those found in granite, are commonly attacked by water that is acid or alkaline in pH. All mineral-formed rocks, except the highly resistant quartz, are broken down and changed in this chemical manner to clay, colloid, silica, granite, and numerous minerals in solution. The common chemical products of weathered rock become the building materials for the sedimentary particles upon which our soil is built.

Besides such chemical weathering of igneous and metamorphic rock, there is mechanical weathering, which is the breakdown of rock into particles without causing change to the minerals in the rock. Heating and cooling by the sun along with water percolating into cracks and pores, separates or expands the rock with enough force to crack off small pieces and disintegrates the rock over time. Thus are the volcanic pressures of soil converted into dust and sedimentary particles to be carried by erosion, wind or water to the flat lands and hillsides where it is labeled *soil*.

24   *Paramagnetism*

Chapter 3
WEEDS

Soon after the eruption, herbaceous vegetation began returning to vacant clear-cuts within the blast zone. The pink flowers of fireweed, and colonizing species of disturbed areas, were a common sight.

Natural History
"Life Returns to Mount St. Helens," Roger del Moral

Less than a year after the eruption of Mount St. Helens the botanist Roger del Moral was working on the sides of the crater. According to his report in the journal, *Natural History*, within a very few months a beautiful weed, known in the far west as fireweed, began to emerge from the ash to "scatter their luminous message of life among the stumps of an earlier clear-cut operation." Fireweed is called fireweed because it is one of the first sturdy plants to emerge from a destroyed clear-cut operation or a forest fire burn. Fireweed is a member of the evening primrose family (*Onagraceae*).

Evening primrose plants are of American origin but were introduced into Europe as one of the European favored wild edibles. It is actually cultivated there for its tasty nut-like flavor. Primrose grows all over North America, from Canada south to Florida, and is a favorite of mine as a flower attractive to the beautiful white-lined sphinx moth. The fireweed is the first cousin to the yellow evening primrose, and in my opinion far more beautiful. Its pink to bright red blossoms cap a bushy clump of lance-head shaped leaves. Framed against a background of grey ash or blackburn, it is a sight fit for people who believe in angels.

I have never observed fireweed in the eastern United States, however it does occur commonly all over the forested regions of the West from Alaska to Mexico. It is also scattered throughout Europe and Asia where land has been disturbed. It is a bright red international signal that man or nature has eliminated the green cover of the land. It is a true forest fire/volcano lover.

Strangely enough, although I never see it in survival books, it is one of the most edible plants. Other names for it are *deer horn*, named for the spike-like prods that stick up beneath the beautiful red, four-petaled flowers, and *wild asparagus* for its tasty, asparagus-like young stems. The tender leaves can also be steamed or boiled as greens. Beekeepers in western states often move their hives close to logged or burned areas where their bees have close access to its sweet honey providing flowers. Why it is called a weed is beyond my comprehension. It should be cultivated as a food plant.

When my daughter, two grandchildren and I took a mule called Blackjack across the High Sierras, Blackjack would stop to browse on a sow thistle or fireweed. They were his two favorite foods and he would pull us off the trail to reach either of them. Mules seem to know more about the beneficial vitamins and trace elements in so-called weeds than modern man — maybe they are smarter in a natural sort of way!

Like Blackjack, the Paiute Indians of California knew all about fireweed and used it also as an anti-diarrhetic and hemostatic herb. I would rename this flower *firefood*.

The real problem is that the term *weed* has come to mean a nuisance plant when, in fact, weeds are without a doubt our most beneficial earthly helpers. The word *weed* should be drummed out of the English language, and weed killers banished from society like all murderers. Nelson Coon, in his classic book on the use of wild plants, calls weeds *wayside plants*. That is indeed a good term for weeds because roadways and trails are disturbed areas and thus ideal spots for the natural cropping of all varieties of weeds. There is not a trail or roadway across the United States that a knowledgeable survival expert could not, as Blackjack did, eat his way along, or even cure a few disorders if so inclined. Other than as a source of food and herbs, the real value of weeds has to do with their soil restorative powers which are awesome. Why is this so?

The short simplified answer is that in depleted soil, weeds send their roots deep down into the more mineral-rich subsoil. In doing so, they pull up into their stems and leaves the very minerals that Julius Hensel speaks about — potassa, sodia, lime, magnesia, manganese, iron, silica, alumina, phosphoric acid, sulphur and fluorine. Weeds are boxed and packaged storehouses of almost every important mineral needed for healthy plant growth. Just as importantly, weeds are natural mixers of such minerals. Since it is done by nature and not DuPont, the magic force called *paramagnetism* is attained in the mixture. It is quite usual in the laboratory to mix such minerals together and attain little or nothing of this paramagnetic force, so why not let the weeds do it correctly?

There are two excellent books on weeds. One is Charles Walters' superb book, *Weeds: Control Without Poisons*, where we find another basis for the book you are reading now. He states: "Plants — weeds as well as crops — actually get about 80% of their nutrition from the air. Most of this nutrition is taken from carbon dioxide and water, but it also includes cosmic and solar energy and airborne nutrients. The effectiveness of this direction of nutrient flow is totally dependent upon two conditions: the inherent integrity of the plant and/or seed and the health of the soil." The health of the soil depends not only on soil nutrients and air nutrients but upon oxygen, as does human life. The atmosphere is composed of only 0.03% carbon dioxide (CO_2) but 20.95% oxygen (O_2). Oxygen is as necessary to root growth deep in the soil as it is to the portion of the plant above the ground which is submerged in the atmosphere. Oxygen, like volcanic rock and dirt, is highly paramagnetic. In fact it is the most paramagnetic of all gasses. If the magnetic force we call paramagnetism is important to the plant above ground, then it must be doubly so in the soil.

The old 1950 classic, *Weeds, Guardians of the Soil,* by Joseph Cocannouer, is a small book that sums up all that weeds accomplish for depleted soil. Mr. Cocannouer lists the four valuable contributions of weeds to the soil. They are:

> *(1) These roots are persistent explorers in a rich world (subsoil) which is to a large degree unknown to domestic crops — until the weed roots build highways leading into it. Thereafter the crops are provided with a more extensive feeding zone. (2) The weed roots pump those "lost" food materials back into the surface soil; (3) the weed roots fiberize the subsoils and (4) help to build a storage reservoir down there for water; water moves up along the outside of the weed roots which feed in the surface layer. That is why a crop on "controlled" weedy land can go through a drought better than a clean crop on similar land.*

I like the term *fiberize the soil*. By fiberize he means the weeds' ability to *eat* their way through compacted soils. Plants, such as weeds, exude from their roots special dissolving substances that soften hard obstructions thus stimulating deep root growth in overly compacted soil. Weeds tend to feed in the lower subsoil zone and will promote an upward movement of capillary water along the outer edge of the root. They are not water robbers as is commonly believed, except in cases where they are too dense in a field. In short, there is an optimal population of weeds among food crops that will enhance the crop growth.

The fiberization of soil by weeds not only allows for water to be pumped upwards, but oxygen to flow in the soil contributing its powerful paramagnetic force to the root growth. Weeds, by aerating the soil, contribute the spongy characteristic that signals fertile soil. Fiberization and aerating of compacted soil allows the roots of the crops in the upper top soil to follow the pathway of the weeds to a better life below. The roots of weeds are tunnelers for the roots of the surface feeding food crops. Mr. Cocannouer makes the point that if one wishes to reclaim eroded and compacted soil, one should even go to the extreme of sowing weed seeds. Heresy!

Before World War II, in Europe and especially in Ireland, one field was allowed to lay fallow for two years before rotating a new crop into it. Weeds were eventually cut and stacked in layers between layers of manure. In the spring the green manure was spread on the row crops. In this way not only were the weeds fiberizing the fallow field, but the nutrients stored in the green weed manure were put back into crop soil.

We may understand then that there are three ways to generate this valuable magnetic force called paramagnetism into the soil:

1. By adding volcanic rock into the soil.

2. By fiberization so that paramagnetic oxygen reaches the roots in soggy soil.

3. By using weeds, which are green containers of paramagnetic minerals, in our compost or manure.

In a future chapter we will cover, in layman's terms, the physics of this mysterious force, how to measure it in your soil, and how to reestablish it on land that is badly eroded and thus depleted of this most necessary antenna for magnetic forces.

32   *Paramagnetism*

Chapter 4
THE NEVER AGING ROCK

Forms which Westerners would consider inanimate have become fused with vitality through Shinto. Whereas we in the West would mould or break natural form to our design, the Japanese, recognizing vitality inherent in the form, shape and design to release the vitality.

The Ocean in the Sand
Mark Holborn

I sat cross-legged at the low table. Outside the darkening and misty raining sky was background to two large, grey colored rocks. One was flat and turtle-shell shaped, the other upright, the "crane and the turtle," I thought to myself. Norio, my Japanese host and good friend, arose to slide open the paper paneled door to the outer enclosed rock garden behind the restaurant.

Norio Inaba invited me to Japan as the main speaker at a four-day symposium in honor of the sacred groves of Ise Shinto Shrine. He was treating me to a traditional Japanese dinner. It was my last night in Ise before leaving for Tokyo. Donald Nordeng, an American friend who spoke Japanese, and Norio were my guides to the countryside and temples of Ise Peninsula.

The Ise Shrine is dedicated to the Sun Goddess and is the Emperor's Shrine — the first and most important shrine in all of Japan. Every twenty years all of the shrines of the complex are torn down and exactly rebuilt in an open space beside each structure. This twenty-year ceremony has been continued since the reign of Emperor Temmu in the seventh century.

The Ise Shrine is actually a complex of two main shrines and many auxiliary shrines. The inner shrine is deep in a sacred forest and the outer shrine closer to the nearby town. Since all of the auxiliary temples are replaced every twenty years it requires at least 80,000 huge logs to rebuild the structures. The logs come from sacred forests deep within the central mountains of Japan. Eighty-thousand logs for temple replacement requires steady growing and healthy trees from healthy soil.

The shrines are beautifully simple wooden structures with elegant thatched roofs. The skill in shaping and refitting these all wooden, and nailless, structures every twenty years is almost unimaginable. It is a telling story of the great love the Japanese people and Shinto priests have for this most beautiful of all the world's nature oriented religions. The history of Japan has been the history of a people who have mirrored the beauty and the forces of nature. Their art is an art of suggestion and not an art of reality. The simple forest temples of Ise suggest the nature gods of forest, rock and sea.

Being the main speaker at a symposium honoring the recently rebuilt Shinto Temples was an honor for a Westerner beyond my wildest expectations. It was as if the nature gods, the *Kami,* or spirits of tree, rock, and soil, had known that 48 years earlier I had journeyed to Ise to visit them, during the occupation of Japan, and they approved of my return. As one Japanese friend said, I was a Catholic/Shinto. But being the main speaker at the symposium was not the only reason I was in Japan. The other reasons were trees, rocks and soil. I was to measure the growing force called paramagnetism in the forest soil. I was also going to study the extremely low frequency (ELF) radio waves, amplified by that soil force, in the atmosphere of the forest.

Paradoxically my earliest indications that rock had a vitality, a living force, came to me climbing cliffs to photograph and obtain falcons and eagles to train. I always had the feeling that falcon eyries were located on very special rocky ledges. I never grew tired while climbing to the rock homes of these magnificent birds of prey. Climbing among the red rocks of Colorado and the granite cliffs of Vermont and New Hampshire, I was intrigued by the fact that the most beautiful and greenest of mountain plants always seemed to grow straight from small cracks in those rocky surfaces. It was as if the rock had a vitality of its own. This is what the Shinto religion teaches, that like plants and animals, rocks are living creatures with a living spirit that the Japanese call *Kami*.

Tree growing from rocky surface.

It was during my war years in Ireland and occupation years in Japan that my belief in the magic vitality, the Kami spirit of rocks, solidified. I can construct in my mind a time sequence of the evolution of my research on this vital force in rock and soil. Most of my year-to-year notes on rock energy were, until about twenty years ago, observational data. In Ireland it was the fact that there was one mountain, Breezy Hill, where I simply felt better whenever I spent a quiet day on its rocky summit.

With the round towers of Ireland, especially the Devinish Tower, it was a fact that the grass was far greener and healthier around the tower than the mainland grass. The Irish themselves attested to this fact, for in those early days they ferried their cattle to the island for grazing —"Aw sure it is the best grass!" Another Irish certitude is that cattle and sheep always gather around ancient stone structures, inside stone rings, and close to megalithic tombs.

In Japan one gets a feeling of restfulness in the wooden and thatch-roofed Shinto shrine of the sacred groves. I began to feel that if the vital force of rocky places made one feel energetic and the wooden shrines and trees of sacred groves made one feel relaxed, that there seemed to be two forces at work. One force was calming and restful, the other energizing and fatigue defeating. Perhaps in Eastern terms, the *yin* of the female and the *yang* of the male?

It was through reading the brilliant writings of the Irish scientific genius John Tyndall that I finally realized that these vital forces were not magnetic — that is, the plus (+) and minus (-) or North and South poles of a magnetic field — but the paramagnetic and diamagnetic properties of rocks and plants. I credit Tyndall because it is important that in describing any new phenomenon, we also make clear where the concept first originated. It is an unfortunate fact of modern of science that there is little credit given for the brilliant, yet simple, early work of such people as John Tyndall. Significant scientific discoveries are usually based on simple experimentation and not million dollar grants. Cancer will be defeated the same way eventually.

Tyndall had been arguing with Faraday about the theory of diamagnetic substances. A diamagnetic substance, especially wood, if placed near a strong magnet by hanging the wood on a string, will be weakly repelled by the magnet. Diamagnetism is a negative movement, or movement away from a magnetic field. Paramagnetism is a strong positive attraction to a magnet. Most organic molecules are diamagnetic and most volcanic rock and ash are paramagnetic.

Tyndall had thoroughly cleaned 30 to 40 species of oak wood from all over the world and was measuring the distance the wood sliver was repelled from the magnet. To his surprise, one wooden splinter was attracted to the magnet indicating that it was paramagnetic. Under the microscope he observed that burned in the wood was part of a company brand indicating the seller of the wood. There appeared to be a very few particles of iron oxide imbedded in the wood. Iron oxide is paramagnetic. The wood was pulled to the magnet even though it was tested at the opposite end from the iron oxide brand. When Tyndall sanded the brand away, the sliver was repelled by the magnet. The few grains of iron oxide had turned the wood from a diamagnetic substance to a paramagnetic substance. Today we would call such a phenomenon, where a small bit makes a big change, solid-state *doping* of the substance. We may understand then that John Tyndall was the world's first solid-state physicist!

It became apparent to me while observing the rock gardens of Japan, especially those in Kyoto, that the placement of rocks in relationship to the sun, as well as the shape of the rocks, were of primary importance. The design principles and secret teachings upon which the gardens of Kyoto rested were apparent in the placement and varying shapes of the boulders scattered about the gardens and parks of Kyoto and the other shrines of Japan.

Interestingly enough, the spacing, and the horizontal and vertical proportions between the principle and subordinate rocks, are often expressed according to the Golden Mean of the ancient Greeks. This tells us in no uncertain terms that antenna design, which is the shape and placement of the rocks operating as antennae in relationship to the sun, is of utmost importance. The Golden Mean is represented by the Fibonacci series (1, 2, 3, 5, 8), where 1 + 2 = 3, 2 + 3 = 5, 3 + 5 = 8, and so on.

It is well known that the science of magnetic dowsing, called geomancy, began in China sometime during China's distant beginnings. Geomancy was the sacred science of placement surveyance. The Chinese utilized a system of divination which is very difficult to understand. That is why it was, and still is, called *Feng Shui* which means wind and water. The method is as fleeting as the wind and as difficult to grasp as the water.

The *Secret Book of Gardening*, published during the Tokugawa period in Japan, contained details of the art of geomancy. According to that tradition, there are male and female stones. The male stone is exposed to the sun and the female stone always lies in a more secluded shaded area. This tells us that geomancy has little to do with north and south pole magnetism, but in actuality, is a paramagnetic/diamagnetic, or sun/shade, phenomenon.

Sometime during the decline of the great Tang dynasty (894 AD) the gates to China were closed by Japan and garden geomancy became garden art, an aesthetic undertaking. It had lost its connection to the sun. Placement and shapes were no longer antenna design, but were more in line with the visions of artists who had little knowledge of low-energy force fields.

There it stagnates today, endlessly fostered by Eastern and Western practitioners who have not the slightest knowledge of the physics of nature's low-energy forces. If one thing should be learned from this book, it is that not only do the laws of physics connect chemistry to biology, but they also connect mankind to the sun!

The Japanese name their rocks. My favorite is the two rock placement called the "turtle and crane." These two rocks represent the mythical Isle of the Immortals, similar to the Western concept of heaven. The taller rock, or crane, is also called the "never aging rock." It symbolizes a young granite-peaked mountain. Below it is the turtle, the "rock of the ten thousand eons," or the ancient turtle of "Horai." These two rocks imply immortality and were the two that I viewed outside the restaurant where Norio and I enjoyed a delicious Japanese dinner.

As I sat cross-legged, trying to manipulate my chopsticks with slightly more grace than my ability apparently allowed, I was struck with the simplicity of the Japanese eating tools. Both are indicative of the fact that, designed by men for life's journey, they do not need to be complex, or even the finished products of energy gobbling machines. They can be as simple as connecting to our environment with wood and stone — a thought to enhance digestion — and also one's vitality.

PART II
THE FORCE

42   *Paramagnetism*

Chapter 5
PARAMAGNETISM & DIAMAGNETISM

In paramagnetism the atoms or molecules of the substance have net orbital or spin magnetic moments that are capable of being aligned in the direction of the applied field.

Dictionary of Chemistry

The above is the technical definition of paramagnetism taken from the *Dictionary of Chemistry*. *The Dictionary of Physics* states the exact same thing in the exact same words! In short, paramagnetism is a physical parameter of all material at the elementary level of atoms and molecules. As the rather trite saying goes, "it shouldn't take a rocket scientist" to figure out that the force is probably just as important to biology as it is to physics and chemistry. After all, physics is what connects chemistry to biology. Despite this irrefutable fact, the word is not found in any of dozens of soil books I have read or even in the *Dictionary of Physical Geography*. It is almost as if God created the paramagnetism force for the benefit of physicists alone so that they could use it to theorize about the makeup of atoms and molecules. I doubt that God so favors physicists, although a goodly number seem to believe it.

Exactly what, in plain language, does this definition of paramagnetism mean? First, we must define the term *magnetic moment*. If you spin a fixed magnet in the center of a loop of wire, you generate electricity in the wire, creating an electric generator (see *Exploring the Spectrum* by the author). Magnetic moment is the ratio between the maximum torque exerted on a magnet or current-carrying coil, or the charge in a magnetic field, and the strength of the field itself. Since atoms and molecules spin, rotate, and vibrate in all kinds of predictable directions depending on their makeup, they are in effect, little dynamic generators displaying both field strength and torque (torque = rotating power in a mechanism). In summary, magnetic moment is the ratio of the strength of the magnetic field to rotating power.

It is obvious that the earth and cosmos itself has a magnetic moment since it is has a low-energy magnetic field of about ½ gauss. *Gauss* is the CGS unit of magnetic flux. *CGS* means Centimeter, Grams, Seconds. Put quite simply, if you have one gram of a substance, one centimeter from a magnet, in what part of one second will it move to the magnet? Put another way, what weight of a paramagnetic material will move one centimeter to a magnet in one second? The Appendix provides tables that list the paramagnetic or diamagnetic properties of some common atoms or molecules.

Any substance, including soil or rock, that will move toward a magnet is paramagnetic. If you can measure the CGS of a substance then you will know the measure of its attractance force to a magnet. CGS is known as *susceptibility* because it is obvious that if a substance moves to a magnet, then it is susceptible to a magnetic field. Other ways to say it are that the substance is attracted to a magnet field, or resonating to the field or grabbed hold of by the field, or even loves the field!

If a paramagnetic substance is placed in a strong magnetic field, all of the field lines will eventually line up, as illustrated:

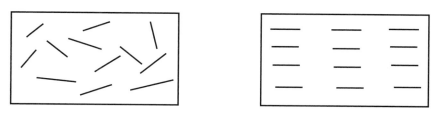

Paramagnetic substance outside & inside magnetic field.

In nature, all substances are in a weak cosmic magnetic field, which is the earth's ever-present ½ gauss, therefore they are aligned thus:

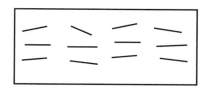

Paramagnetic substance in weak magnetic field, e.g., earth field.

They are then not completely random, or, as mathematicians might say, in a complete chaotic arrangement. That is why chaotic mathematics is so important to a study of paramagnetism. Take heed chaotic mathematicians. Once placed in a strong magnetic field like the electromagnetic coil of a CGS meter, they become more aligned.

The measure of the more aligned is the measure of the paramagnetic force, or the CGS measure.

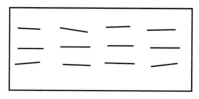

Paramagnetic substance becomes even more aligned in the magnetic field of a CGS meter.

Now that we know that paramagnetism is the alignment of a force field in one direction by a substance in a magnetic field, then we must ask, what is *diamagnetism?* The *Dictionary of Chemistry* defines diamagnetism as follows: "Diamagnetism is the magnetization in the opposite direction to that of the applied magnetic field, *e.g.*, the susceptibility is negative away from the magnetic field." Actually all substances are diamagnetic, but it is a weak form of magnetism and may be masked by other, stronger forces, for instance a magnetic field.

Diamagnetism results from changes induced in the torque by bits of electrons that oppose the applied magnetic flux. There is thus a weak negative susceptibility to the magnet. Most organic compounds, including all plants, are diamagnetic. If plants are diamagnetic and good growing soil paramagnetic, then we must be dealing with the *yin* and *yang* of Chinese and Japanese geomancy, or the energy put forth by the crane and turtle rock formation.

Why are the crane and turtle rock important? Simply because most of the ancient Zen gardens that I have observed over the years appeared to be both paramagnetic/crane and diamagnetic/turtle! This was observed and documented in the *Secret Book of Gardening*. The diamagnetic properties of the flattened turtle rock are visually obvious by the amount of white quartz in it. One does not chip pieces of beautiful Zen garden rock to study its CGS properties, but most quartz is not only recognizable by sight, it is also either neutral or weakly diamagnetic.

The Nanzen-en stroll garden of the Kamakura period has several high granite and low quartz boulder arrangements as does the Ryogen-in garden designed by Soami. The diamagnetic/paramagnetic, or yin/yang arrangement is most often seen in the double crane and turtle configurations. There is also a triple configuration that has a central granite standing rock and two smaller granite paramagnetic lower rocks.

Tentoku-en, the landscape garden of the Momoyama period, has a high crane basalt rock and low turtle limestone rock. Around these rocks an arrangement of Chinese bellflowers grows in profusion. Interestingly enough, they grow to the left of the tall basalt crane rock and on the right side of the flatter turtle rock.

By positioning such rocks in relationship to the sun and to each other, one can control plant growth. Apparently the ancients knew about this yin and yang, diamagnetic/paramagnetic phenomenon and utilized it in their Zen gardens. That such knowledge is now lost is demonstrated by the fact that the crane/turtle arrangement found at the elegant restaurant where my friend and I had dinner was composed of stones that were both paramagnetic and not paramagnetic/diamagnetic.

Before we move on to a discussion of atmospheric ELF radio waves, it is important that we also define magnetism (ferromagnetism). *Ferro* means iron. Magnetism occurs in ferromagnetic substances because it is a characteristic of certain metals, particularly iron, at certain temperatures. Below a certain temperature, called the Curie point, an increasing magnetic field applied to iron, or any ferromagnetic substance, will cause increasing magnetization to a value so high that it becomes saturated and remains permanent. In short, a magnet is a metal that has a permanently stored, aligned magnetic moment. It is analogous to a stored DC battery.

Magnetic substances are extremely rare in nature, the best known being the mineral magnetite. Because of the rarity of magnetite, it is not apt to be the growing force of nature. That does not mean that magnetism is unimportant in the scheme of life.

In this regard, there is one last point that should be made. Even though magnetism is a fixed force, it does vary slightly. There is no such thing as flat line DC — everything in nature alternates, at least slightly. The simple fact is that the magnetic field of the cosmos and the earth alternates far more than the field of a fixed DC magnet. It is this alternating earth/ cosmic field to which volcanic soil and volcanic rock resonate, or to which both are susceptible.

As in the case of plants, water is diamagnetic. The atmosphere, because of the oxygen, is paramagnetic. Some of my preliminary experiments at night, during the full moon, indicate a paramagnetic/diamagnetic, plant, moon, water and soil relationship in nature. We know that the moon, which is highly paramagnetic, has a very strong effect on tides, which are of diamagnetic water. The many volcanic and/or meteorite cones indicate a paramagnetic moon body even though I could find no data on this subject from moon rock measurements.

It has long been known that certain Indian tribes planted by the full moon. There is little doubt in my mind that the American Indian knew more about good agriculture techniques than modern agriculturists! As the Sioux brave remarked while watching a farmer turning under virgin prairie grass, "wrong side up!" (in *Altars of Unknown Stone*, by Wes Jackson).

Chapter 6
ELF ATMOSPHERIC RADIO

My dear son:

    You are interested in radio-telephony and want me to explain it to you. I'll do so in the shortest and easiest way which I can devise. The explanation will be the simplest which I can give and still make possible for you to build and operate your own set and to understand the operation of the large commercial sets to which you will listen.

Letter of a Radio-Engineer to His Son
John Miles

The above book was written in 1922, the year before I was born, which proves that the invention of radio-broadcast by the dentist Mahlon Loomis was well on the way when I came along in 1923. My friends and readers who know me will be surprised by the name Mahlon Loomis being substituted for Nikola Tesla. However, when Loomis, an Illinois dentist, first sent messages between two wire antenna from two Blue Ridge Mountains in Virginia (Bears Den and Catoctin Mountains), Nikola Tesla was just eight years old. He had hardly invented anything at that age.

During those early years of electrical studies, dentists were using static generator machines to produce radio emission in the 2,000-Hz (Hz = cycles per second) region. It is a region of the ELF spectrum that anesthetizes one's nerves. It was very useful for pulling teeth, but later the AMA stopped research on electrical anesthesia for one simple reason: drugs, though unsafe, make more money. That region of the electromagnetic photon spectrum is still found on early spectral charts — it has been eliminated from later charts. Out of mind, out of use!

Dr. Loomis was born July 21, 1826 in Oppenheim, New York. He studied dentistry and taught school in Cleveland, Ohio. He received a first patent for Koolin artificial teeth. He married Achsah Ashley of West Springfield, Massachusetts in 1856. After five or six years of experimenting with his static spark machine, he built a unique radio aerial transmitter. In 1866 he gave the first public demonstration of his system. His notebook description of his wireless is dated February 20, 1864. It reads:

> *Two kites were let up — one from each summit — eighteen or twenty miles apart. These kites had each a piece of fine copper wire gauze about 15 inches square attached to their under side equipments and apparatus of both stations were exactly alike. The time piece of both parties having been set exactly alike. It was arranged that at precisely such an hour and minute the galvanometer at one station should be attached, or be in circuit with the grounded and kite wires. At the opposite station the ground wire being already fast to the galvanometer, three separate and deliberate half-minute connections were made with the kite wire and instruments. This defected, or moved, the needle at the other station with the same vigor and precision as if it had been attached to an ordinary battery. After a lapse of five minutes, as previously arranged, the same performance was repeated with the same result until the third time. Then 15 minutes precisely were allowed to*

elapse. During which time the instrument at the first sta-
tion was put in circuit with both wires while the opposite
one was detached from its upper wire, thus reversing the
arrangements at each station. At the expiration of the
fifteen minutes the message or signals came to the initial
station, a perfect duplicate of those sent from it, as by pre-
vious arrangement. And although no transmitting key
was made use of nor any sounder key to voice the messages,
yet they were just as precise and distinct as any that ever
sped over a wire. A solemn feeling seemed to be impressed
upon those who witnessed the performance as if some grave
mystery hovered there around that simple scene, notwith-
standing the results were confidently expected.

Mahlon Loomis, the Discoverer
and Inventor of Radio, *Otis B. Young*

On July 30, 1872 the United States issued its first patent in radio (No. 129,917). It was titled "Improvement in Telegraphing" and went to Dr. Loomis. It was not, of course, a wire telegraph system, but rather a wireless radio system. Since there was no battery, as noted in his scientific notebook, why did it work? That question is the basis for the fact that Dr. Loomis' rather simple experiment is one of the great experimental scientific works of all times. It demonstrates that Mahlon Loomis understood quite well the atmospheric waves that ninety years later were called Schumann waves, named after the German Scientist W.O. Schumann who theorized them in 1954.

The fact that the system was 600 feet of wire attached to a simple non-battery galvanometer is certain proof that the photon and electrical energy was generated in the sky between ionosphere and ground. Since his experiment between mountains in Virginia was on a clear day, he certainly understood that the atmosphere contained electronic charges from distant lightning to power his system. His last paragraph demonstrates that he was both scientist and poet.

Unfortunately, as with most really great discoveries, Dr. Loomis was denied credit for discovering radio communications by a smug and greedy group of lobby types. Twice he applied for $50,000 in funding to develop his research and twice was refused by the Senate. Actually, the Loomis Aerial Telegraphic Company was eventually chartered by Congress which authorized a Capital stock of $200,000 with the privilege of an increase to $2,000,000. The Senate passed the bill and President U.S. Grant signed it.

Loomis continued to experiment in West Virginia with wooden towers and steel pipes. He never received his money and died at his brother's home in Terra Alto, West Virginia. Buried in the old local cemetery, in near anonymity, lies the real inventor of radio broadcast.

I cannot determine that a single electrical scientist or written history of radio, such as *Syntony and Spark — The Origins of Radio*, by Hugh G.J. Atken, ever spoke of or quoted this brilliant scientist. Intellectual dishonesty has always been more rampant in science than elsewhere. If I, an entomologist, know of Loomis, then surely most American electrical engineers and, in particular, physicists must have known of him. Their lack of credit to him diminishes all in my view.

The so-called "Schumann" waves, which would better be called atmospheric "brain waves," occur in the 8- to 30- Hz region. Since they exactly match the human brain waves (8, 14, 21, 27 and 33 Hz) I consider that the EEG brain waves are actually the low-frequency atmospheric waves "in the air" of our brain. We are, after all, mostly water and air. The organic molecules of our bodies are only little photonic oscillators that fill the spaces between the water and atmosphere of our body. It is highly unlikely that the exact match between low-frequency atmospheric ELF (8, 14, 21, 27 and 33 Hz) and so-called brain waves is accidental.

I suggest that this change in our paradigm regarding brain waves will infuriate the elitists of science who utilize mathematics to obscure what is obvious in nature. Mathematics is a beautiful and elegant subject which should be utilized to design systems based on nature, not utilized by high-tech technocrats to confuse the public with computer models that have no relationship to what is observable in nature.

In summary, every living human being is like a sponge in a bowl of jelly (the atmosphere). When the atmospheric jelly shakes (ELF waves), then the jelly in the sponge also shakes at the same frequency. The organic photon oscillators of our body superimpose their messages on this atmospheric brain/body continuum (think about ESP).

Somewhere in the world there are at least 2,000 to 4,000 lightning bolts per minute. These keep the atmosphere lit up in the radio region of the spectrum between the ionosphere and the paramagnetic soil of the ground. We may consider the system analogous to a fluorescent light bulb where an electric spark, at either end of the tube, keeps the mercury vapor lit up in the UV region, which in turn visibly lights up the white chemical in the bulb called phosphor.

Since the covering of the ionosphere around the earth creates a giant resonant cavity, resonant standing waves of energy in the ELF (extremely low frequency) and VLF (very low frequency) radio region occur. I have been able, with my detector, to detect the brain wave region (8, 14, 21, 27, 33 Hz), the electrical anesthesia region (600 to 4,000 Hz), and the lightning region (25,000 to 50,000 Hz).

My patented detector for measuring these frequencies is a piece of jute, or burlap, cloth soaked in seawater. NASA and Penn State utilize a huge long coil of wire, somewhat similar to what Dr. Loomis utilized to pick up low-frequency atmo-

Photonic Ionic Cloth Radio Amplifier Maser (PICRAM).

spheric waves. I had to design a much cheaper and lighter system so it could be carried to the top of trees to measure ELF and VLF energy in the forest canopy of the Amazon.

Trees, because they are excellent dielectric antennae, resonate very well to ELF and VLF frequencies and even the higher broadcast frequencies. Everyone knows that if you grab the antenna of your little transistorized receiver the sound comes in stronger. That is because your body is an excellent antenna/amplifier in the radio portion of the spectrum. Trees, because of their size, are even better antennae. In India, rural villagers utilize trees for radio and TV reception.

Like Dr. Loomis' long wires from flying kites, barbed wire fences and power grids, which may be thousands of miles long, are excellent antennae for ELF and VLF atmospheric waves. Nearby fences and power grids show both atmospheric waves and 60 cycles.

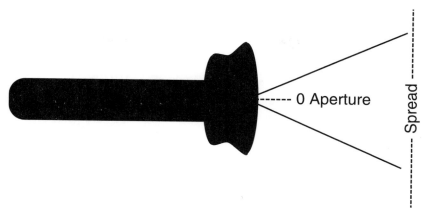

Light spread.

Since ELF/VLF frequencies saturate all of the atmosphere, they are not going to my cloth antenna from point A to point B (as between broadcast tower and receiver), but rather the cloth detector is imbedded in the atmospheric frequency. Therefore the energy is at what is called *zero aperture* by antenna engineers. Thus, there is no *spread* of the radio *light* as from a flashlight. Since there is no spread, total spatial coherence is *at* the cloth detector. In other words, the radio is *spatially* coherent, not necessarily *time* coherent. Coherent means all the energy is in one space at one time like marching soldiers, not a mob scattered, or incoherent.

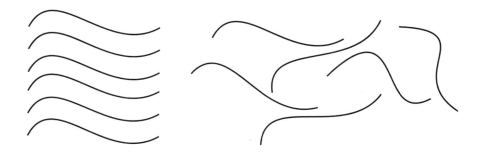

Coherent lines, left, and incoherent lines, right.

In order to make absolutely certain that my ELF/VLF tree frequencies were not artifacts of 60 Hz (or 50 Hz in Europe and Ireland), I created checks by taking measurements in sheltered valleys miles from power stations. I was never satisfied with my U.S./European checks, so I flew down to the Amazon in Peru where there is still 25,000 square miles of jungle with no power plants. It is the land of the Aschuara tribe of so-called headhunters. My friends Kenneth Silver, Paul Beaver and I were the first white persons ever to go up the Huagramona River, or River of the Tapers.

The so-called headhunters were a gentle people, very much like the Irish. They guided us and climbed to the canopy with my detector and oscilloscope. The frequencies, though weaker, were very much present, proving them to be part of the atmosphere, amplified by both trees and the highly paramagnetic soil, which acts as one plate of a huge condenser for the ELF/VLF radio radiation.

When researchers from Oxford and Arizona radio carbon dated the Shroud of Turin they did not utilize a check for comparison as I did in my experiments. In short, they were either incompetent scientists, or the work was a predetermined con job, take your pick.

My work with these important natural frequencies began over 50 years ago with my first visit to an ELF/VLF antenna amplifier called a round tower in Ireland. Then, of course, I was using my own "dreamy" brain for an antenna! That is a deadly statement for a scientist and will no doubt diminish my reputation among high-energy technocrats.

Chapter 7
ROUND TOWERS

It cannot be too strongly emphasized that written records on this subject are all very much secondary. There is no contemporary account of the building of a round tower nor of its purpose.

The Round Towers of Ireland
George Lennox Barrow

Sometime after VE day, I took my jeep about twenty miles or more from Belleek, Ireland to Enniskillen, a beautiful town nestled on a narrow strip of land between Upper and Lower Lough Erne. In Enniskillen I searched out a good natured Irish fisherman and rented his time and boat to row me out to Devinish Island. On the way across the choppy waters of Lough Erne, he brought me up to date on the magic of that ancient monastery island. He pointed out that the local farmers shared its grazing rights and ferried their cattle to its shores since the grass was the finest in the country. The great taste of grass-fed Irish beef is no secret.

That day, like D-Day which was not so long before my island visit, will remain forever in my mind. I remember it as being a black, stormy, rainy day, so wild that even the great

Catalina flying boats and white swans stayed close to the shore.

That dismal day should have been enough to dampen anyone's spirits. Instead, from the moment I landed on the island, it was as if I had entered another world.

Shipping orders to Ireland in World War II always read: *cold, wet, and windy*, and Devinish lived up to that description. Despite the storm, I remained on the island until after dark when my fisherman friend returned for me. The entire day was one of euphoria — a dream world magnified. I have since returned to Devinish five times and visited most of Ireland's other remaining, or partially remaining, sixty-five round towers.

Photograph of Devinish round tower, Ireland.

Author's World War II drawing of round tower.

Round towers are unique to Ireland alone and are arranged on the ground in a recursive pattern in relation to the stars of the night sky above.

In later years I began to wonder, exactly what was the magic of these towers? No one seemed to really understand why they were built, certainly not as hiding places for monks to escape Vikings or bandit attacks as many books speculated. Nobody in their right mind would run into a smokestack to escape an attacking enemy.

Except for the Scattery Island tower, at the mouth of the River Shannon, all round tower doors are from nine to fifteen feet above the ground. This is what led to the simplistic view that they were places of refuge — as if Vikings could not climb or light a fire to smoke victims out. Why then, I ask myself, were round tower doors nine to fifteen feet above the ground?

Half of science is in asking the right questions. The right question was: "Could the towers be some form of dielectric radio antenna for focusing lightning-radio waves?"

In Southeast Asia and the Philippines I had often noticed bamboo shoots growing during electric storms as if viewed by time-lapse photography (see *A Walk In The Sun* by the author). Lightning-radio waves, in those early days, were called static or noise. In radio there is an old saying, "One man's noise is another man's signal." Lightning static was obviously my signal if it could speed up bamboo shoot growth.

Since round towers are not metal, as are most low-frequency radio systems, then they had to be dielectric, waveguide antennae for photon energy. A photon is a mathematical particle of energy that describes the behavior of the spectrum from radio waves at one end to gamma radiation at the other (see *Exploring The Spectrum* by the author).

In 1953, D.G. Kiely wrote a small book called *Dielectric Aerials*. That book was my bible for studying the waxy spines, called *sensilla*, on insect antenna. Dielectric resonators are what later became known as fiber optics waveguides. They guide and amplify electromagnetic waves. A dielectric is an insulative substance that can be a semiconductor, a substance that weakly conducts current.

Our friend John Tyndall discovered dielectric waveguides in water when he shone a light down a flow of water from a hose. He noticed that the light followed the water without spreading out to the sides. It makes John Tyndall not only the first solid state physicist, but also the first photonic waveguide, or fiber optic, scientist. This is another simple, and great, experiment that anyone can replicate by holding a waterproof flashlight and shining it up from the bottom of a faucet flow to observe the light In the case of a transparent tube like plexiglass, a water dielectric or fiber optics, the light

Flashlight experiment illustrating photonic waveguide effect.

goes up the center. Kiely's book covered both ordinary wave-guides but also what is called open resonators. I discovered that insect sensilla are open resonators, that is, the energy travels on the surface of the spine and not down the center. Since insect spines are in the micrometer range in length (1 micrometer = 1/1,000 of a millimeter) then the wavelength must be of the dimensions (or multiples thereof) of the spine length.

Something unique occurred while I was studying the sensilla under the microscope. They were, as often as not, small

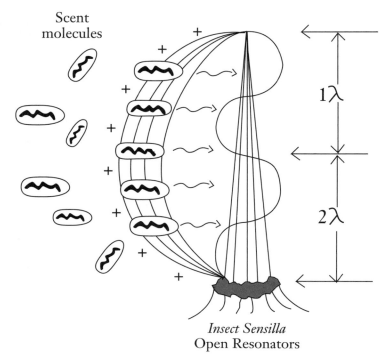

Scent
molecules

1λ

2λ

Insect Sensilla
Open Resonators

Insect sensilla.

micrometer-long models of the huge religious structures I had
observed in my walk around the world. If insect antenna sensilla resonate to infrared frequencies from vibrating molecules
such as scent, why couldn't stone rings, megalithic
tombs, round towers or great Gothic Cathedrals be dielectric
antenna-waveguides for ELF radio waves? They are the
dimensions of the much longer radio waves.

Early in the 1970s, I carried my old D.C. 222 Tektronix
oscilloscope to Ireland. By means of the probe and a loop of
seawater-soaked jute cord, I connected it to the base of the
tower at Glendalough. The round tower at Glendalough is in
perfect condition, and sits at the head of a beautiful wooded
valley about thirty miles south of Dublin. It is my favorite
tower.

Religious structures around the world function as dielectric waveguides like the insect sensilla shown at top right.

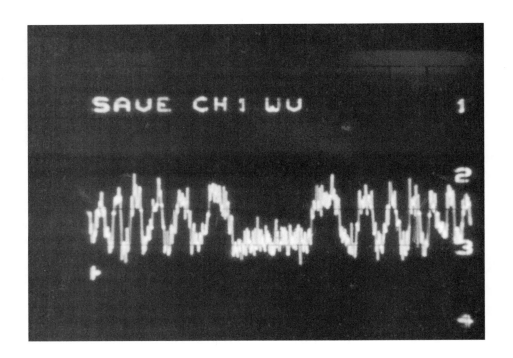

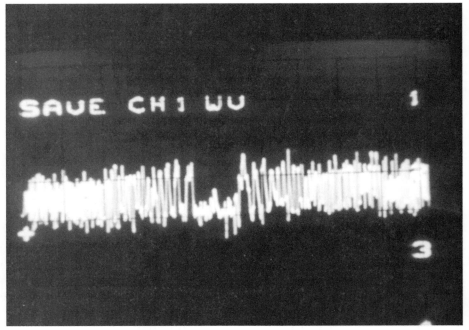

Side view of target waves on my oscilloscope.

Impedance is a form of electrical resistance to current flow between two parts of an electronic circuit. The impedance match between the wet cord and the stone tower was much better than I had ever thought it might be. I was in for a delightful surprise. As Dr. Loomis said, "A solemn feeling seemed to be impressed upon those who witnessed the performance as if some grave mystery hovered there around the simple scene, notwithstanding the results were confidently expected." Unlike Dr. Loomis, the results I obtained were not confidently expected — they were only slightly hoped for! The crowd around me was not a group of friends but tourists.

As I sat at the base of the beautiful Glendalough tower trying to shield the scope face from the sunlight, I began to notice waves coming in with a very peculiar pattern. I soon realized that I was looking at what, when seen at high energies, physicists call *target waves,* because they resemble a bow and arrow or rifle target with a large center circled by narrow outer rings. There was one difference, I was looking at target waves from the side.

Another difference was that the side-viewed outer waves were not evenly spaced as in a regular target wave, but varied from narrow rings to rings that got wider and wider as they passed across the scope. Furthermore, they came from a series passing in one or two seconds to one I measured at Dog Rock in Australia that lasted four hours.

In the old days (1946-48), when I was installing 300-KHz radio range stations in Japan during the occupation, we usually had to construct a false ground of wire mesh six to ten feet above the real ground to keep the radio beams stable. Sometimes, over heavy clay soil, the real ground was stable enough, like a metal plate of a condenser, to maintain a strong signal with no artificial ground. In every case, however, it was necessary to raise the tower base on an insulator

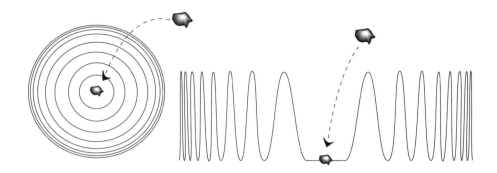

Target waves from above and side views can be seen when a rock is thrown into water.

six to ten feet up in the air. This is because there is always a null or low signal right at the ground due to the fact that the wave is reflected back from the ground and cancels itself at ground level. The tower wave also has a "shadow" wave in the ground making a half-wave antenna in reality a full-wave antenna.

I measured the Glendalough tower at ground level — no signal! I slowly raised my saltwater-soaked cord up the tower. At six inches the signal was weak, at three feet it began to come in quite strong. As I approached the door which faces south-southeast and is 3.20 meters above ground level, the signal increased in strength until it was 20 mV, at its strongest on the scope, right at the bottom of the doorway.

The tower is constructed of mica-schist and granite, both paramagnetic. The mortar of round towers is believed to be made with ox blood, also making it paramagnetic.

I spent the rest of the day measuring and plotting ELF energy around Glendlough round tower. In every measurement, the atmospheric ELF at 8 Hz, 2,000 Hz, and target wave region (from 300 Hz down to 0 at the middle and back up to 300), wherever the detector touched the tower, increased in amplitude from three to eight times.

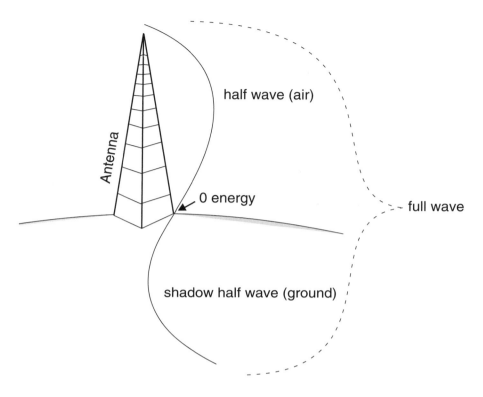

Radio tower and shadow half wave.

I had discovered that round towers are indeed high tower ELF radio antenna paramagnetic amplifiers. More astonishing yet, I discovered the ancient Irish monks of the 5th to 9th centuries were rock antenna radio engineers.

Most round towers of Ireland are now without floors or ladders, however, I was able to visit the tower on Scattery Island where the door is at ground level. Inside I discovered that without even touching my jute-saltwater cord to the walls, that at the center there was a two- to four-time increase in the strength of the waves.

The 8-Hz and 2,000-Hz waves always came in strongest at dawn and dusk. I recorded the same ELF phenomenon in so-called megalithic tombs and even found a megalithic pictograph of a target wave on the side of the chamber at Loughcrew.

It is not possible for a modern "plastic" technocrat to even begin to understand my feelings as I traveled back in my mind to the very days those kindly Irish Catholic monks were caring for their people and their lands. The best proof of what I say is the propaganda for clear cutting which assures that the round tower, paramagnetic soil ELF force will wash to the sea. It is certain famine, as is very close in Russia now, and certain death for our great-grandchildren. It is up to the American farmers to save us urban greedy from ourselves and of course, perhaps the Cape May warblers also!

Chapter 8
THE CGS METER

Fertilizer production and other human activities have more than doubled the global rate of nitrogen fixation since preindustrial times. The resulting imbalance is contributing to ecosystem disruption, ozone depletion, greenhouse effects and other environmental problems.

Human Impacts on the Nitrogen Cycle
Ann P. King and Robert H. Socolow

The above statement is the summary heading of the lead article in a special issue of *Physics Today* on "Physics and the Environment," published in November 1994. It is satisfying that an elitist university, Princeton, has finally realized the fact that chemical farming is killing mankind, Cape May warblers, and — I would venture to say — earthworms and honeybees! One cannot help but wonder what the solutions coming from such institutions will be.

Most university funding now comes from large grants made by corporations or from government agencies which are controlled by large corporations through PACs and lobbyists. Corporate control of university research is a phenomenon that slowly developed from university agricultural experiment stations' acceptance of money from chemical companies in exchange for insecticide testing. I am an authority on that subject. In the fifties and sixties I tested just about every insecticide known to mankind at the time.

As corporate conglomerates take over more and more in a desperate attempt to make life linear, we are developing a form of dictatorship I term "corporate communism." Two excellent books on the demise of the scholastic university are *Prof Scam*, by Charles J. Sykes, and *The Closing of the American Mind*, by Allan Bloom.

Why do I ask what will happen when the common-sense research by Princeton researchers begins to yield solid results? Quite simply, it is because today's intellectual atmosphere is infused by a desire to control nature. Put more simply, modern researchers tend to practice manipulation and not science. Science is discovery, manipulation is the belief that one can utilize discovery to shape nature to man's will. That, of course, cannot be accomplished, for nature's plan is God's plan.

One does not need to read this book to come to the realization that manipulation is today synonymous with science. Almost every newspaper has articles on man's unique ability to manipulate DNA in order to create "better", more "vigorous, or "more money yielding" plants. DNA research lends itself to this man-is-God syndrome. Materialistic skeptics will maintain that farmers manipulate nature. That is very true of chemical farming where one imposes on nature what are, in essence, poisons to eliminate problems.

Eco-agriculture is not manipulation, but simply replacement agriculture. It is the ancient, or Irish-Celtic if you prefer, form of agriculture where mankind worked very hard to replace what he took from the land. Crop rotation is one example of this sort of farming.

Man was originally a hunter/gatherer. Somewhere along the way he discovered that he could not only gather animals together for food by herding, but he could also gather plants by field farming. This form of endeavor is not manipulation, it is simply a method for the concentration of food resources for the purpose of a more efficient gathering, which we call the harvest.

When I went to school at Arkansas and Kansas State, the agricultural experiment stations encouraged the scientific study of efficient gathering. Dr. Reginald Painter, with whom I worked at Kansas State, was the world's most renowned gatherer. He went out into the fields of Kansas, Oklahoma and Nebraska and found individual wheat plants that were resistant to disease and insect damage. He brought them back to the laboratory and in his greenhouses worked on breeding, by natural means, large populations of these naturally resistant varieties.

My job was to discover why they were resistant. In particular, I studied corn resistance to the corn earworm moth. I soon discovered that if I grew my small, 1/8-acre plots on poor soil they attracted far more corn earworm moth eggs than plots on dark, well aerated, bottomland soil. In other words, healthy plants require healthy soil. And healthy plants, like healthy people, simply do not attract disease or insects. Why?

It took me forty years to find out why and that is the subject of a much longer, more complex, scientific paper that will be of little use to the family farmer trying to grow food for my family and others who like to eat. Put quite simply, it is because unhealthy plants from "sick," poison-fed, soil give off slightly higher ethanol and ammonia infrared signals than healthy plants. This is particularly true of modern farmed ammonia-drugged plants. Ask any professional entomologist what are two of the most universal attractants of insects, and they will agree they are ethanol and ammonia, both precursors of fermentation and death. Farm crops should be harvested before they reach old age, the attractive state that brings hoards of nature's scavengers, disease and insects, to feed on them.

We have all heard of the "Halls of Ivy." Why does ivy love old red-bricked walls? Why do weeds grow out of deserted concrete roads or plants from old stone castles? The 18th-century drawings of Irish round towers show them covered with plant life right up to the top windows. After World War II most of them were cleared of plant growth and are now protected by the Irish Department of Public Works, as they should be. It is the paramagnetic force in brick, stone, and concrete that stimulates such growth. Plants do not take hold in diamagnetic limestone walls.

If crops are grown on soil that is well aerated, composted, and filled with those little critters that stabilize it but fail to grow well, attracting insects and disease, then what is missing in such soil is probably that mysterious magnetic force called paramagnetism. I call it mysterious for several reasons, most particularly from studying the values given for diamagnetic and paramagnetic substances in the *Handbook of Chemistry and Physics, 50th Ed.* We find, for instance, that oxygen (O_2) is highly paramagnetic (+3,449), while mercury (Hg) is diamagnetic (-24.1) Yet when they combine to form mercuric oxide (Hg_2O), one gets a diamagnetic reading of -76.3. Take another nice combination: calcium (Ca) is paramagnetic (+40) while nitrogen (N_2) is diamagnetic (-12) whereas calcium nitrate $Ca(NO_3)_2$ is diamagnetic (-45.9). In other words, +40 + -12 = -45.9!

The CGS world is indeed very strange, or perhaps I should say, confusing. Nowhere in any physics book have I ever found a really good explanation of the combining powers of diamagnetic atoms and molecules with paramagnetic ones. There are a few educated guesses but they actually explain little.

I have always considered it unfortunate that most scientific CGS meters, like the Bartington Model MS2, cost between $4,000 and $8,000, depending on attachments. Originally I measured CGS by hanging soil tubes or rocks from a thread and measured how far one gram moved to a 2,000-gauss magnet in one second. This takes hours of repeating measurements, and yields only comparisons between low and high values (see Appendix). My work was originally published in the book *Ancient Mysteries, Modern Visions*, the first study of the phenomenon in agriculture or soil, I might add.

The word *paramagnetic* is brandished around like seagulls in an ocean gale, but still little understood by the average scientist or agricultural consultant, much less the farmer. Hopefully, this book will be a partial remedy.

The Bartington type CGS meter works on a single chamber principle, and will measure from 1.999 to 1 million CGS. The meter's single chamber is referenced against 1-ml cylindrical vial of H_2O which has a CGS standard of 0.719. Distilled deionized water is utilized. When a sample material, like a rock, is placed within the influence of a low-frequency alternating magnetic field, produced by the chamber sensor coil, a change in the coil frequently results. The new alternating frequency is converted to a magnetic susceptibility (CGS) reading and displayed on a digital meter.

All volcanic soil and rock is highly paramagnetic, giving a CGS from 200 to 2,000 CGS. Good soil is therefore highly paramagnetic. What is needed for the practical working farmer is a simple hand-held meter that will read positive paramagnetism from 0 to 2,000. Such a meter has been designed by myself and Lee Leitner, a consulting electrical engineer, and Dr. Edward O'Brien, professor of electrical Engineering at Mercer University in Macon, Georgia. With the help of Bob Pike, we have spent over a year perfecting this instrument and have named it the P.C. Soil Meter (PCSM) which could mean the Paramagnetic Count Soil Meter or the Phil Callahan Soil Meter. The choice is yours! I have tested it all over the United States and in Japan and Australia. It is rugged and works to perfection. It is not calibrated down to three decimal places, as is Bartington's research meter, but reads on a digital scale from -100 up to 2,000 CGS.

The author has demonstrated that good healthy crops grow only on highly paramagnetic soil. The paramagnetic soil meter (PCSM) is based on an entirely different principle than the Bartington's meter, and compares the sample to the paramagnetic (CGS) properties of the atmospheric oxygen which is the most paramagnetic of all gasses, CGS 3,449. The principle was first delineated by Philip Callahan, Edward O'Brien and Lee Leitner, and the meter was designed around this two-coil chamber principle.

The PCSM can be sold for a cost estimate of $400 per unit, a cost savings of thousands of dollars, and is thus accessible to family farmers who could not afford a $4,000-$6,000 meter. Because it is designed for farming and not research purposes, it is not calibrated down to many decimal places. It reads out in increments with a simple digital meter where 0 to 100 CGS = poor; 100 to 300 CGS = good; 300 to 800 CGS = very good; and 800 to 1,200 CGS and above = excellent. The higher the CGS, the healthier and better the crops. Excellent soil comes from paramagnetic volcanic soil. All really good soil is volcanic. This force can be added to soil, where it has eroded away, by spreading ground-up paramagnetic rock such as basalt, or granite, into the soil.

The two-chamber system is based on the fact that the air space of both chambers is filled with atmosphere alone, therefore any soil sample or rock added to the sample chamber will unbalance the impedance match ($X_L.X_C=0$) so that the unbalanced side will read the mismatch as a figure converted on the meter to a CGS. This is a simple procedure based on the same exact electrical principle as matching the coil-tuned circuit of a transmitter to a resonant antenna system, where the transmission line connecting transmitter antenna must also be in resonance ($X_L.X_C=0$), that is 0 reactance or resistance. Since the PCSM meter is solid state and the sample chamber of the same size as a plastic holder for a 35mm film cannister, it is light and portable in the field.

With this meter many a farmer can save his soil from the destruction advocated by present day chemical propaganda. This author estimates that 60 to 70% of this volcanic paramagnetic force has been eroded away worldwide. A meter to measure this force is therefore absolutely necessary in order to save our chemically raped and eroded soil. Soil should be alive with living organisms such as bacteria and earthworms, diamagnetic plant material such as compost, and the rich soil paramagnetic force. Mineralization of the soil by adding separate minerals does not, I repeat *does not*, mean that the paramagnetic force has been added. We know little of the effect of living forces in rock, but we do not install into the soil the living paramagnetic force by blind mineralization of the soil.

Complex mixtures do not necessarily contain a high CGS/paramagnetic factor. It is an absolute fact that all chemical fertilizers measured by this inventor, even those labeled organic, impart or contain the CGS force at such a low reading as to be totally useless in reinforcing the remaining natural volcanic force of the soil.

I have never asked a person to follow my advice without some sort of common sense method to test it. That is what real science, not manipulation, is all about. The farmer that measures this force, or follows my simple flower pot experiments (see Appendix), is just as much a scientist as a university-type who uses the fine Bartington instrument to get down to the last decimal place. We scientists love decimals.

The last sections of this book will give tables with details of measurement by the PCSM meter, and simple experiments using flower pots that will demonstrate the beauty and elegance of God's paramagnetic force. May you proceed with love of your soil from here.

84 *Paramagnetism*

Epilogue
ACCUSE NOT NATURE

Accuse not Nature, she has done her part
Do thou but thine, and be not diffident
of wisdom, she deserts thee not, if thou
Dismiss not her.

Paradise Lost
John Milton

In his marvelous book called *Healing Sounds: The Power of Harmonics*, Jonathan Goldman discusses the power of sound harmonics and relates an experience of a lifetime that he was a part of at the ancient Mayan pyramids at Palenque. His Mexican guide took him, along with five companions, into the subterranean chamber below one temple that was ordinarily closed to tourists. The guide pointed his flashlight toward one section of the chamber and asked Jonathan to chant his harmonic voice toward the wall. The guide switched off his flashlight leaving the group in the total blackness of a cave. As Jonathan chanted the beautiful harmonics of his voice in the darkened granite-limestone chamber, it lit up as if by magic. The group of Americans were astounded, for they could actually see each other in the glow of the sound-stimulated light.

Goldman describes it as a more subtle phenomenon than the light from a flashlight. He was astute enough to point out that he had created light from sound. He also realized that it was definitely not the phenomenon, postulated by physicists, of a sound wave being speeded up to the point that it turns into light — an accomplishment that is not likely to happen in the near future, but will eventually be achieved by science.

Lighting up the atmosphere with sound is a phenomenon I have been easily able to accomplish since 1969. My work on insects irrefutably proves the mechanism of light, or infrared, generation by sound harmonics. The narrow band coherent infrared radiation from insect scent molecules is stimulated to emit by an electric sound wave (E wave) in the ELF radio region. See my rubber band, radio wave experiment in *Exploring the Spectrum*. A snapped rubber band gives off an ELF radio wave in the 100- to 200-Hz region.

Certain musical sounds have long been known to stimulate plant growth. The hum of a bee is not for the benefit of human poets but for the benefit of the bee. It stimulates the various scents that control the behavior of the bee.

By stimulating a mixture of ethanol and ammonia, or sex scents, with an ELF wave, I can create photon harmonics throughout the infrared, and even visible, regions of the spectrum. The electric sound field vibrates the molecules in the same manner that plucking a rubber band vibrates the air and produces sound and radio ELF emission. Those harmonic waves travel up and down the photon electromagnetic spectrum the same way harmonics do from a plucked banjo string. In one direction, the harmonics move to higher frequencies along the infrared region and even move into the visible region. When such waves hit solid objects like cells or insect antenna, they scatter out other waves of electromagnetic energy that are actually coherent (marching together) and amplified. Pictured is a Fourier transform spectrum (original recordings for historical purposes) of just such waves. The experiment shows cabbage looper sex scent modulated by ELF

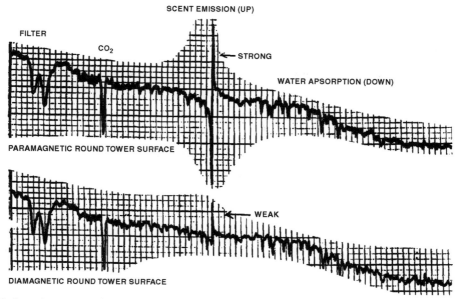

Infrared emission from scent scattered against two different round tower surfaces (see text).

(60-120 Hz) lab visible light and blown out at five mph wind speed across two model round towers.

At the far left, the spectrum shows a wide absorption line from the polyethylene filter protecting the clean spectrophotometer mirrors from the heavy molecular cabbage looper sex scent. Next, it clearly shows the carbon dioxide (CO_2) absorption line (14.9 μm) from CO_2 in the air space between the mirrors. At the far right is a series of weak, room water vapor infrared lines. All of these absorption frequencies represent the same principle as a radio receiver receiving radio radiation from a radio station.

In the center of the graph one can observe a continuous absorbing in the downward line. The upward line shows emission in the narrow-band infrared line from the cabbage looper female sex scent, also called a pheromone. It is stimulated to emit the coherent, or partially coherent, line by the 60-Hz

ELF from the room light bulbs as it is blown across and collides with the rough surfaces of two model round towers.

Note that the sex scent blown and scattered off the carborundum paramagnetic tower surface has five times the height (power) of that scattered against the bismuth diamagnetic round tower surface. Think of the paramagnetic tower surface as vibrating at the very weak magnetic resonance of the scent molecule and thus pushing the scent molecules away from itself with more energy than the opposite, and much weaker, vibrating diamagnetic round tower.

We may easily understand then that the paramagnetic forces of rock amplify not only ELF radio waves in the atmosphere generated by lightning, but also the photon waves generated in the infrared and visible control region of life. Life processes are electronic, like the nervous system, but also very much photonic. Life's complex communication system's messages are carried by photons, as are AT&T's.

All through this book I have talked of CGS as a measurement of the subtle and marvelous weak paramagnetic force. What I have left unsaid, until the end, is that all of these numbers should be multiplied by 10^{-6} or $1/1,000,000$ of a CGS. In other words, any of our CGS measurements is actually one millionth of a CGS — an extremely weak magnetic field, for example, 90 CGS actually means $90/1,000,000$ of a CGS. The very words *paramagnetism* and *diamagnetism* imply very, very weak magnetic fields.

Weak fields, that are difficult to measure and work over long periods of time, are the antithesis of modern high-energy, nature-destructive science because they imply God knows more about life than the FDA or the Academy of Science. A Nobel Laureate, Irving Langmuir, points out that there is no such thing as low-energy systems, for example, cold fusion, and we lesser mortals should bow to his superior knowledge.

There are many elegant and brilliant women scientists today. One of the most brilliant in the field of brain research is Dr. Beverly Rubek, director of the Center for Frontier Sciences at Temple University. I quote Dr. Rubek directly as she put it far better than I, proving once again that women have more common sense than men:

> *Irving Langmuir, awarded the Nobel Prize in Chemistry for his work in surface chemistry, also explored what he called "pathological science." Although he did not publish on the latter, he presented a lecture on the topic in 1953 that was transcribed and published in* Speculations in Science and Technology *in 1985, and in* Physics Today, *October 1989, pp. 36-48. This famous talk is often taken as the last word on what constitutes "sick science." It lumps phenomena such as N-rays, mitogenic rays, the Allison effect, extrasensory perception, and more into the dumpster. Langmuir claims that there is nothing underlying these effects whatsoever, and that their proponents are simply "true believers" who are mistaken.*
>
> *Beyond downright dishonesty on the part of their proponents, Langmuir believes that all these phenomena have certain features in common. The characteristic symptoms of "pathological science" according to him are as follows: (1) the maximum effect observed is produced by a causative agent of barely detectable intensity, and the effect is independent of the intensity of the cause; (2) observations of the effect are near the threshold of human observations; (3) there are claims of great accuracy; (4) there are fantastic theories proposed to explain the effects that are contrary to experience; (5) criticisms are met by ad hoc excuses; and (6) the ratio of the supporters to the critics rises up somewhere near 50% and then falls gradually to oblivion, as the critics cannot reproduce the effects.*
>
> *While I would certainly agree that not all phenomena examined by scientists are "real" or important, I question these as the features distinguishing good from bad science.*

In fact, I would argue that several of these features do not stand up at all under examination. For example, Robert Millikan, who first measured the charge on the electron in his famous oil drop experiment, actually found partial units of charge on occasion that he apparently ignored because they did not agree with the larger body of data. Yet it was these fractional charges together with other evidence that led to the concept of quarks comprising the electron and other so-called elementary atomic particles. Had Millikan paid attention to such occasional phenomena, he might have gone further than elucidating features of the electron.

A similar story is true for genetics. Mendelian or classical genetics is never exactly obeyed in cross-breeding experiments. It was precisely such small quirks in the patterns of inheritance demonstrated in maize genetics that led Barbara McClintock to refine our concepts about DNA in her Nobel Prize-winning work on the translocation of genes.

Thus, it would seem that phenomena that are often at the threshold of detectability, which are the exceptions to the rule that challenge our notions of ordinary experience, and which may require radically new theories to explain them, sometimes lead to important scientific breakthroughs.

Appendix I
Tables & Figures

Tables

Among all of the rocks I tested, there are very few that read high for the paramagnetic force. That is why the PC Soil Meter is so necessary to agriculture and good plant growth. Note that basalt rock may vary from a low of 73.4 (Table III) to as high as +470 (Table V). As Table II demonstrates, there are very few strong gasses except oxygen, or strong metals or other paramagnetic molecules. Granite may vary all the way from +5 (Table V) to 1,769 (Table VI). Note that almost all apparition sites, as well as Indian or ancient healing sites, are highly paramagnetic. The ancients and the Virgin Mary knew all about ELF paramagnetic forces.

Table I
IRISH ROCK MAGNETIC FORCES*

Type	Magnetism	Force (1 to 5)
Basalt	Para	+5
Green Slate	Para	+4
Old Red Sandstone	Para	+/-3
Chalcopyrite	Para	+3
Connemara Marble	Para	+/-3
White Granite	Para	+2
Carboniferous Limestone	Para	+2/-1
Dolomite (Limestone)	Para or Dia	+1 or -1
Quartz	Para or Dia	+1 or -1
Sphaherite	Para or Dia	+1 or -1
Barite (Barium Sulfate)	Para or Dia	+1 or -1
Galena (Lead Sulfide)	Dia	-1

*The force is my own measurement in millimeters (mm) distance that 1 gram of paramagnetic rock will move to a 2,000-gauss magnet. Certain rock types may be para- or diamagnetic depending on where they originate. Limestone is mainly diamagnetic. Diamagnetism is an extremely weak force and never moves away from a 2000 gauss magnet more than one mm (-1). We see from this table that the green slate of Yeat's Ballylee Castle in Ireland is extremely paramagnetic. From *Nature's Silent Music* by the author (*Acres U.S.A.*, 1992). Some stone, such as marble and limestone, may be either slightly paramagnetic or diamagnetic (+ or -).

Table II
MAGNETIC SUSCEPTIBILITY OF METALS, GASES AND OTHER SUBSTANCES*

METALS	*CGS Reading (10^{-6})*	
Iron (Fe)	Ferro (magnetic)	
Cobalt (Co)	Ferro (magnetic)	
Nickel (Ni)	Ferro (magnetic)	
Aluminum (Al)	Paramagnetic	+16.5
Copper (Cu)	Diamagnetic	-5.4
Zinc (Zm)	Diamagnetic	-11.9
Silver (Ag)	Diamagnetic	-19.5
Lead (Pb)	Diamagnetic	-23.0
Mercury (Hg)	Diamagnetic	-24.1
Gold (Au)	Diamagnetic	-28.0
Antimony (Sb)	Diamagnetic	-99.0
Bismuth (Bi)	Diamagnetic	-280.1

GASES

Oxygen (O_2)	Paramagnetic	+3,449.0
Helium (He)	Diamagnetic	-1.8
Hydrogen (H_2)	Diamagnetic	-3.9
Neon (Ne)	Diamagnetic	-6.7
Nitrogen (N_2)	Diamagnetic	-12.0
Argon (A)	Diamagnetic	-19.6
Carbon Dioxide (CO_2)	Diamagnetic	-21.0
Chlorine (Cl_2)	Diamagnetic	-40.5
Bromine (Br_2)	Diamagnetic	-56.4

Table II (continued)
MAGNETIC SUSCEPTIBILITY OF METALS, GASES AND OTHER SUBSTANCES

OTHERS	*CGS Reading (10^{-6})*	
Calcium (Ca)	Paramagnetic	+40.0
Potassium (K)	Paramagnetic	+20.0
Sodium (Na)	Paramagnetic	+16.0
Selenium (Se)	Diamagnetic	-25.0
Silicon (Si)	Diamagnetic	-3.9
Surphur (S)		
forms a, b, 1	Diamagnetic	-15.5
		-14.9
		-15.4
form g	Paramagnetic	+464 to +700
Water (H_2O)	Diamagnetic	-13.1

Handbook and Chemistry and Physics, 50th ed.

Table III
ARIZONA ROCKS*

Rock or Mineral	Weight/grams	CGS (5 Readings)
Paramagnetic		
Magnetite	6	+ 4,428.4
Hematite	4	+ 467.2
Petrified Wood	9	+ 354.6
Basalt	16	+ 73.4
Garnet Schist	3	+ 9.0
Epidote	2	+ 2.0
Thulite	4	+ 1.0
Diamagnetic		
Andesite	1	0 to -.001
Coal	1	0 to -.001
Talk	2	0 to -.001
Flourite	7	0 to -.001
Pyrite	4	0 to -.001
Opal	2	0 to -.001
Dolomite	5	0 to -.001
Barite	2	0 to -.001
Onyx	2	0 to -.001
Quartz	2	0 to -.001

*Identified by geologists

Table IV
KANSAS ROCKS*

Rock or Mineral	Weight/gram	CGS (5 readings)
Shale	11	+ 18
Sphalerite	11	+ 17
Opaline sandstone	13	+ 15
Red sandstone	15	+ 13
Bentonite clay	11	+ 9
Red shale	10	+ 7
White volcanic ash	10	+ 3
Bituminous clay	14	+ 3
Sand and gravel	12	+ 2
Plastic fire clay	5	+ 1
Chalk	12	+ 1
Oil shale	8	-.003
Halite (salt)	6	-.001
Chert	5	-.001
Anhydrite	11	-.001
Kansas limestone	10	-.001
Gypsum	7	-.001
Asphalt rock	9	-.001
Diatomaceous mare	2	-.001
Galena (lead ore)	3	-.001

*Identified by geologists

Table V
IRISH ROCKS*

Rock or Mineral	Weight/gm	CGS (5 readings)
Basalt	12	+470
White granite	32	+ 5
Dolomite	10	+ 5
Barite	23	+ 5
Slate	8	+ 5
Green Marble	10	+ 3
Old red sandstone	32	+ 2
Chalcopyrite	23	+ 1
Carboniferous limestone	7	+ 1
Galena	53	+ 0
Quartz	45	-.001
Sphalerite	35	-.001

*Identified by geologists

Table VI
SACRED PLACES

PLACE

Apparition Sites of the Virgin Mary	*CGS (5 Readings)*
Tepeyac Hill, Mexico	+840 rock
Le Puy, France	+1,020 rock
Medjugorje, Bosnia-Herzegovina	+ 700 soil
Kerrytown, Ireland	+1,769 rock
La Salette, France	+ 580 rock
Loudres, France	-1 to +3 rock*

Indian and Ancient Structures

Kings Canyon, Australia	+4,795 rock
Ayres Rock, Australia	+30 rock
Dog Rock, Australia	+780 rock
Huagramona River, Peru	+560 soil
Hueco Tanks, Texas	+361 rock
Bagmore Stone Circle, Ireland	+247 rock

*Soil covering limestone grotto may be highly paramagentic but I could not obtain a reading.

FLOWER POT FARM EXPERIMENT

Take two plastic flower pots. Fill both with potting soil from the same bag. One pot should be left plain. In the other pot, place a paramagnetic stone or sandpaper model of a round tower (15 to 60, proportion of diameter to height) and place it in the middle of a plastic (non-paramagnetic) flower pot. Take a pack of garden radish seeds and plant them 1/4 to 1/2 inch deep, about 3 or 4 seeds per hole, around the pots. Water each day with the exact same measured amount of water. After eight days of 70-80° growing temperature, pull them up and weigh the root's "held in place" soil. The astonishing results demonstrate plant control by the paramagnetic force. Note how the roots and soil mimics the energy force pattern of a man-made radio station (based on weight).

Please note, I do not ask my reader to believe what I say, but I do ask them to see for themselves.

Author's note: Two high school students in science fairs won local and state awards for experiments based on this work utilizing paramagnetism.

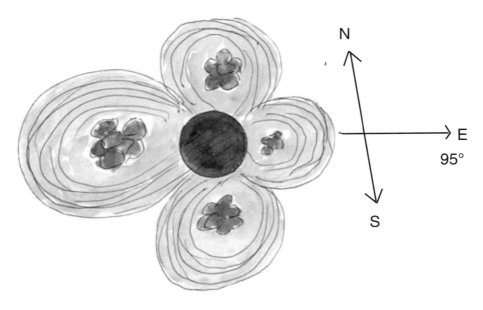

N

E
95°

S

Round tower planting.

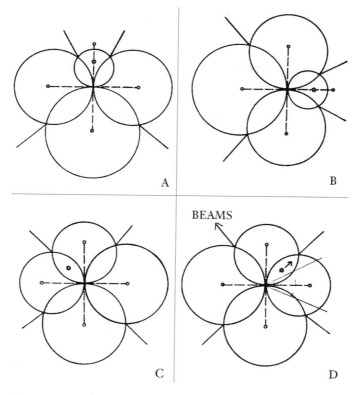

BEAMS

Belleek radio range. Patterns secured due to presence of course-bending antennae.

The ELF growth pattern force of energy focused into the ground by the paramagnetic soil, round towers, or rock can be easily plotted by planting radish seeds around the rock, round tower, or in soil mixed with ground up rock.

In this red sandstone tower example it will be noted that the tower is oriented with th door facing 95° east toward the rising sun in mid-September in Gainesville, Florida. In such a system, the least energy is to the east resulting in slow growth and small plant size and the greatest energy is to the west producing fast growth and large plant size. Side growth is intermediate. Such a plot based on plant size and root-dirt weight at an eight day harvest, is very similar to plots of energy from my World War II radio range station in Belleek.

Roots pulled out of potting soil.

The largest root growth, with the most fine rootlets, is at top left to the west of the round tower. The smallest is at 95° east at the lower right of the photo. The north growth at the top right is slightly smaller than the south growth at bottom left.

Planting around rocks.

Top photo, growth at 12 days; bottom photo, growth at 16 days. Note from compass that the strongest growth is to the east at 95°, in contrast to tower and tomb plantings. The higher growth rate and root complex is always off the sharp corners of such highly paramagnetic rocks. I first noticed this growth effect while climbing cliffs and searching rock canyons for eagle and falcon nests as a youth.

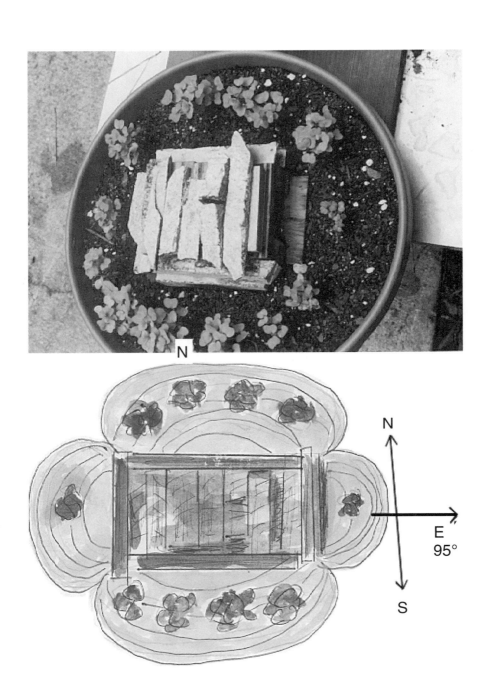

Plot of model megalithic tomb.

108   *Paramagnetism*

Note energy is weak at front entrance and strong along the sides and rear. This model is of a Vermont megalithic stone structure. Constructed of diamagnetic wood interior and paramagnetic pink granite exterior.

It appears that most healing/religious structures such as gothic cathedrals, round towers, and megalithic tombs are facing east so that the weak energy is at the entrance and the strong energy is at the back where the altar of healing chamber is located. There is also stronger energy at the sides, where the arms of the tomb cross the main tunnel as seen in gothic cathedrals.

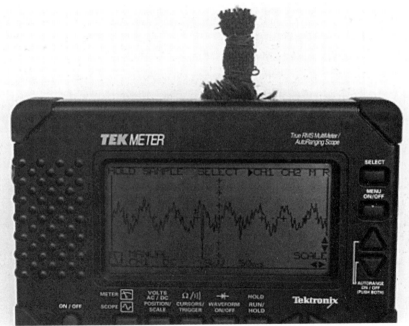

The PICRAM system.

PICRAM, Photonic Ionic Cloth Radio Amplifier Maser, is my name for the patent (No. 5,247,933) I obtained for my ELF (extremely low frequency) antenna detector. As you can see from the bottom photo, it is mounted directly on the Tek meter oscilloscope input with no lead. On the 5 mV range, it accurately measures ELF atmospheric waves generated by lightning which are detectable even underground in soil. These waves stimulate plant root growth.

The PICRAM is constructed by soaking wool-linen cloth or burlap in seawater. In the top photograph, the cloth is connected to a simple banana plug at the corner and wrapped around the plastic of the plug where it is held in place by two rubber bands.

Harry Kornburg, my patent co-author, translated the Hebrew which describes such a piece of cloth worn by the Jewish High Priest. It enhanced his immune system in order that he could safely examine lepers like those sent to him by Christ. The bible is by far the best science book for low energy systems ever written. The Hebrew name for my PICRAM ELF detector is Shatnez. It was worn as a long ribbon strap wrapped around the high priest's body. (See Appendix II for copy of patent).

GLOSSARY

Amplitude — Strength of an electromagnetic wave, usually shows as the height of the wave on an oscilloscope.

Antenna — A metal or dielectric (insulative) substance from which electromagnetic waves (photons) are transmitted or received.

Aperture — The cross-beam diameter of a focused beam of light or other photon energy; also the effective diameter of a lens or mirror.

C.G.S. — Centimeter/grams/second — the measurement of the magnetic flux density (gauss), see text for explanation.

Coherent (radiation) — Electromagnetic radiation in which two or more waves travel in unison which peaks and troughs together.

Clast — A fragment in sedimentary rock.

Cleavage — The breaking of some minerals long one or more regular directions.

Diamagnetic — Magnetization in the opposite direction to the applied field, *e.g.*, away from a magnet. All plant life is diamagnetic.

Dielectric — A nonconductor of electric charges that under certain conditions can be a semiconductor, insulative substance.

Dike — An intrusive thick vertical sheet of igneous rock which cuts across other types of rock.

Doping — mixing a minute amount of one substance into a large volume of liquid, gas or solid.

ELF-VLF — Extremely low frequency (radio) from 1 Hz to 10,000 Hz for ELF, and 10,000 Hz to the broadcast band for VLF. See the definition of spectrum in author's *Exploring the Spectrum*.

Erosion — The disintegration of rock due to wind, water, or soil, or the movement of the soil.

Fault — A fracture of the earth's crust.

Fiberization — the ability of the roots of weeds to penetrate compacted soil with the aid of secreted chemicals.

Field — A region in which a body experiences a force as the results of the presence of some other body or bodies.

Fold — The plastic deformation of rock strata.

Form — Shape of a mineral or rock.

Fracture — Random way which minerals break.

Gauss — The CGS unit of magnetic flux density. See CGS.

Geode — A rock concentration, rounded and often hollow which contains crystals.

Harmonic — An oscillation having a frequency that is a simple multiple of a fundamental oscillation.

Hertz — The unit of frequency equal to one cycle per second. Symbol Hz.

Igneous rock — Rocks formed when volcanic magma solidifies, for example, basalt.

Incoherent — Frequency (cycles) that are out of step, non-resonant.

Kami — Japanese term for the spirit of an object, rock tree, etc.

Lava — Molten rock reaching the earth's surface where it rapidly cools. Fine grained.

Magma — Molten rock beneath the surface of the earth.

Magnet — A piece of magnetic material, e.g. iron, nickel or cobalt, that has been magnetized and is therefore surrounded by a magnetic field.

Mantle — Zone between the earth's crust and iron-nickel core.

Metamorphic rock — Rock that has been altered from its original form by heat and pressure at great depths.

Oolite — Spherical grains of calcite that builds into limestone.

Paramagnetism — The atoms or molecules of a substance that have a net orbital or spin magnetic moment and are capable of being aligned in the direction of the applied field. (see text).

Photonic — Pertaining to photons. A photon may be regarded as a unit of radiation (light) energy. It is a mathematical concept.

Radio Wave — A wave for transmitting information in which the medium is long wave electromagnetic energy (above ELF-VLF region).

Schist — Type of metamorphic rock with layerings of mica.

Schuman Waves — Long wave extremely low frequencies (ELF) radio waves in the 8, 14, 21, 27 and 33 Hz region. They are generated by lighting in the atmosphere.

Sedimentary rock — Rocks formed by the accumulation of sediments from weathering and erosion.

Sensilla — The minute spines on insects of many diverse shapes that resonate to infrared frequencies from oscillating molecules.

Sill — A horizontal sheet of intrusive rock injected between layers of sedimentary or metamorphic rock.

Susceptibility — A magnetic term for the ability of a substance, e.g. rock, to receive and transmit magnetic fields from the cosmos.

Vein — A more or less upright deposit of minerals that cut through other rock.

Volcano — A vent in the earth's crust through which magma, gasses, and volcanic ash are ejected.

Waveguide — A substance that guides radiation, *e.g.* light, radio, etc., along its axis with very little energy loss.

Yang — The male (+) force in Chinese lore.

Yin — The female (-) force in Chinese lore.

Appendix II
The Patent

United States Patent [*]

Callahan et al.

Patent Number: 5,247,933

Date of Patent: Sep. 28, 1993

PHOTONIC IONIC CLOTH RADIO AMPLIFIER

Inventors: Philip S. Callahan, 206 NW 27th St.; Harry Kornberg, 1352 NW 61st Terr., both of Gainesville, Fla. 32605

Appl. No.:	**772,652**
Filed:	**Oct. 7, 1991**
Int. Cl.	**AB1B 5/00**
U.S. Cl.	128/639
Field of Search	128/399, 400, 379, 849 128/403, 889, 635, 639, 734

References Cited

U.S. PATENT DOCUMENTS

407,673	7/1889	Mellon	361/224
497,822	5/1893	Royer	361/224
871,479	11/1907	Cooper	361/224
3,867,939	2/1975	Moore et al.	128/400
3,916,447	11/1975	Thompson	128/849
3,997,785	12/1976	Callahan	250/338
4,205,685	6/1980	Yoshida	128/399
4,834,079	5/1989	Benckhuijsen	128/379

OTHER PUBLICATIONS

Senior, *The elusive Search for Electroanesthesia*, Medical Instrumentations, vol. 18, No. 1, pp. 86-87, (1984).

Callahan, *Studies on the Shootborer Hysipyia grandella (Zeller) (Lep., Pyraliadae), XIX, The Antenna of Insects as an Electromagnetic Sensory Organ*, Turrialba, vol. 23, No. 3, pp. 263-274, (1973).

Callahan, *Moth and Candle: The Candle Flame as a Sexual Mimic of the Coded Infrared Wavelengths from a Moth Sex Scent (Pheromone)*, Applied Optics, vol. 16, No. 12, pp. 3089-3096, (1977).

Callahan et al., *Mechanism of Attraction of the Lovebug, Plecia neartica, to Southern Highways: Further Evidence for the IR-Dielectric Waveguide Theory of Insect Olfaction*, Applied Optics, vol. 24, No. 8, pp. 1088-1093, (1985).

Callahan, *Dielectric Waveguide Modeling at 3.0 cm of the Antenna Sensille of the Lovebug, Plecia*

neartica Hardy, Applied Optics, vol. 24, No. 8, pp. 1094-1097, (1985).

Callahan, *Nonlinear Infrared Coherent Radiation as an Energy Coupling Mechanism in Living Systems*, Molecular and Biological Physics of Living Systems, pp. 239-273, (1990).

Primary Examiner — **Kyle L. Howell**
Assistant Examiner — **George Manuel**
Attorney, Agent, or Firm — **Sterne, Kessler, Goldstein & Fox**

ABSTRACT

A Method and apparatus for detecting radio waves that propagate along the atmospheric boundary layer of human skin. This function is realized with the use of a photonic cloth constructed of flax and wool, soaked in a saline solution and air dried, and subsequently placed upon the human skin. The radio waves can then be monitored by connecting the photonic cloth via a set of probes to an oscilloscope.

12 Claims, 5 Drawing Sheets

PHOTONIC IONIC CLOTH RADIO AMPLIFIER

Background of the Invention

1. Field of the Invention

The present invention relates generally to a method and apparatus for detecting the radio frequencies that propagate along the atmospheric boundary layer of human skin.

2. Discussion of Related Art

The present invention described herein is based on the early work of Snape, d'Arsonval, Robinovitch, and Leduc. In particular, Snape pioneered the use of extremely low radio frequencies (ELF) as an anesthetic in dental extraction (Snape, J., *On electricity as an anesthetic in dental extractions*, Trans. Odont. Soc. Gr. Brit., pp. 287-312. (1869)). Subsequently, in 1890, Arsine d'Arsonval demonstrated that ELF pulsed electrical currents, ranging from 2500 Hz to 10,000 Hz, induced general anesthesia in humans. Similarly, in 1902, Leduc demonstrated that a pulsed electrical DC current applied to the central nervous system could effectively induce anesthesia. Robinovitch did extensive work in the area of electric analgesia sleep and resuscitations

---

*The oscilloscope photos (Figures) are not included in this copy of my patent. An example of the most important target (healing) wave is shown on page 68.

Appendix II 117

(Robinovitch, L.G., *Electric Analgesia Sleep and Resuscitation Anesthesia* (chap XVI), ed. J.T. Gwatheny. D. Appleton & Co., New York, pp. 628-643 (1914)). More recently, Czaja demonstrated that treatment in the ELF frequency range enhance the immune system (Czaja, W., *Comparative Studies of Electroanalgesia and Barbiturates*, Polski Archivum Weterynaryjne, pp. 205-224 (1986)).

Between 1965 and 1973 applicant demonstrated that antennae sensilla on insects act as photonic waveguides to collect and transmit infrared frequencies. From this early research, applicant postulated that living systems (e.g., insect spines and plant fibers) also utilize the radio portion of the frequency spectrum to energize photons from radio and infrared emitting molecules. The requirement for detecting and or stimulating infrared and radio emissions from living systems is the ELF modulation of the organic and gaseous interface located at the waxy surface of the system. That is, living systems store coherent photon emissions from the external environment which become part of the self-organization of the living system. It has been demonstrated that ELF frequencies in living systems range from 10^3 Hz in nerve action potentials to 10^{-2} Hz for physiological functions.

From this prior research, applicant has determined that radio waves in the ELF region of the radio spectrum are propagated along the atmospheric boundary layer of the human skin. ELF in the range of 800 Hz to 5200 Hz averaging 1000 Hz, with narrowband 10,000 Hz to 150,000 Hz sideband ELF radio signals are natural to the skin surface. The 700 Hz to 10,000 Hz region of the frequency spectrum is the region of so called radio "whistlers" (i.e. radio signals) from atmospheric lightning strikes around the world. It is this atmospheric electricity that modulates the frequencies from the atmospheric boundary layer of the skin. These modulation frequencies are equivalent to the 3 Hz to 10 Hz oscillations discovered by Schumann stimulated by lightning. These flicker modulations (which are approximately 3 Hz to 6 Hz) can be observed on an oscilloscope while measuring the 1000 Hz and 10,000 sidebands present on human skin.

FIGS. 1,2 and 3 of the appended drawings are readings of an oscilloscope showing the radio signals in the 700 Hz to 10,000 Hz portion of the ELF radio spectrum that are emitted from normal, healthy human skin. These signals were detected by touching the oscilloscope probe to the photonic ionic cloth radio amplifier and touching the face of a cathode ray tube with the hand. A battery (DC) operated 222 Tektronix hand held digital storage oscilloscope and capacitance coupling, with no AC interference, was used for detecting these frequencies in this manner. At a 5 mV range and a 1 mS

sweep time the amplitude ranges from ½ mV (weak signal) to 30 Mv (strong signal).

The oscilloscope sweep shown in FIG 1 has approximately two main 1000 Hz frequencies (between approximately 800-1200 Hz), shown at C_1 and D_1, which are 180° out of phase and occur at exactly 8.4 ms apart. At high amplitudes the two main broad band frequencies generate a series of narrow sidebands of approximately 10,000 Hz, shown in FIG 1 between A_1 and B_1. The 10,000 Hz sidebands are emitted when the two main 1000 Hz frequencies reach an amplitude of 15 Mv or higher. As shown in FIG. 2, there may be as few as one sideband, as shown at A_2, to as many as fifteen sidebands. At extremely high amplitudes there is a main band frequency splitting. As few as one sideband to as many as eight sidebands emit from the region of the 1000 Hz signal under such high amplitude conditions. FIG. 3 shows an example of an oscilloscope sweep at an extremely high amplitude, having two sidebands, shown at A_3 and B_3.

BRIEF DESCRIPTION OF THE INVENTION

These and other advances concerning electricity and its effect upon living systems, as well as the discovery that radio waves in the ELF region are propagated along the atmospheric boundary layer of the human skin are utilized by the present invention. The present invention includes a method and apparatus for detecting the radio frequencies that propagate along the atmospheric boundary layer of the human skin. This function is realized with the use of a photonic cloth constructed of flax and wool which is soaked in a saline solution and air dried, and subsequently placed upon the human skin. When the photonic cloth is placed in contact with the the skin in has an electroanesthesic effect on the body.

BRIEF DESCRIPTION OF THE DRAWING AND SPECTRUM

The foregoing and other objects, features and advantages of the inventions will be apparent from the following more particular description of the preferred embodiments of the invention, as illustrated in the accompanying drawings in which:

FIG 1. is an oscilloscope recording showing the ELF radio signals that are emitted from normal human skin;

FIG 2. is an oscilloscope recording showing the potential for ELF radio frequencies at extremely high amplitudes to have a single sideband.

FIG 3. is an oscilloscope recording showing the ELF radio signals that are emitted from normal human skin;

FIG 4. is a perspective diagram of a woven photonic cloth of the present invention;

FIG 5. is an oscilloscope recording taken from a piece of woven photonic cloth soaked in saline solution;

FIG 6. is a magnification of the recording shown in FIG 5;

FIG 7. is a magnification of the recording shown in FIG 6 showing the details of the first 1000 Hz frequency;

FIG 8. is a magnification of the recording shown in FIG 6 showing the details of the second 1000 Hz frequency;

FIG 9. is an oscilloscope recording taken from a 6 inch by 15 inch woven photonic cloth with the right hand of a lab assistant held approximately one foot from the cloth and left thumb capacitance coupled to an oscilloscope;

FIG. 10 is an oscilloscope recording taken from a 2 inch by 6 inch knitted photonic cloth with the left hand of a lab assistant touching the cloth and the right thumb capacitance coupled to the oscilloscope;

DESCRIPTION OF THE PREFERRED EMBODIMENTS

As shown in FIGS. 1, 2 and 3, radio waves in the ELF region are propagated along the atmospheric boundary layer of the human skin. In particular, 1000 Hz (between approximately 800 Hz to 5200 Hz) and narrowband 10,000 Hz to 150,000 Hz sideband ELF radio signals are natural to the surface of the skin of the human body. The narrow sidebands vary from person to person (e.g., due to the health of the person), time of day and weather conditions, although the 1000 Hz and 10,000 Hz sidebands are continuously emitted from the skin's atmospheric boundary layer. The highest peak of these emissions occurs at dawn and dusk (i.e. between 0630 to 0930 hrs. and from 1830 to 2130 hrs.) The 1000 Hz and 10,000 Hz sideband frequencies can be detected, and amplified, by the photonic ionic cloth of the present invention.

Referring to FIG 4, one embodiment of the photonic cloth of the present invention is shown generally at 400. Photonic cloth 400 is constructed as a plain weave, comprising warp yarns 405 made of flax and weft or filling yarns 410 made of wool. Each warp yarn is a single yarn, while each filling yarn consists of three smaller yarns combined to form a single yarn. The flax used to form yarns 405 is natural and untreated (i.e., *Linum usitatissimum*). Similarly, weft yarns 410 should be made from natural, untreated wool. Thus, both the flax and wool should be unblended and unwashed so that the lanolin remains in the wool, and the waxy outer layer remains on the flax. The natural flax acts as a dielectric waveguide (i.e. it is photonic) due to its waxy characteristics. Although one yarn of flax is sufficient as warp yarns 405, experiments have shown that two or more yarns in combination will also detect and generate the 1000 Hz and 10,000 Hz sideband frequencies.

The photonic cloth can be any size. However, in the preferred embodiment of the present invention, cloth 400 is approximately two inch by six inch to six inch by fifteen inch. In the alternative, the cloth could be woven as a belt approximately three inch by forty-eight inch long.

In addition to weaving, the cloth could be knitted using any known technique utilizing natural and unblended flax as the warp yarns and natural and unblended wool for the filling.

To enhance the ability of the cloth to stimulate and/or detect the radio emissions from the skin surface, it is soaked in a saline solution for approximately one to six hours and then air dried until it is just slightly damp. The saline solution preferably consists of an isotonic aqueous solution containing a borate buffer system and sodium chloride, preserved with 0.1% of sorbic acid and disodium (EDTA). An alternative is to use four tablespoons of sea salt per ½ pint of water with the same borate buffer as described above. Ocean or seawater could also be used. The saline content in the damp cloth acts as an ionic detector for the radio energy emitted from the human skin. In particular, the hollow and fanshaped (i.e., branched) wool fibers act as an insulator, storing and feeding moisture to the waxy flax which absorbs the salt and thus becomes a photonic waveguide detector. Furthermore, the wool acts as a condenser by keeping the system electrically charged above what it would be charged if the cloth was made of saline treated flax alone. Thus, the cloth should be kept slightly damp during use. In order to maintain this slight dampness, the cloth may be placed between two polyethylene layers or their equivalent and sealed to retain the slight moisture. It is important to maintain the cloth in a slightly damp condition, because if the cloth is completely dry or very damp the cloth will not function properly.

Turning now to FIG. 5, an oscilloscope recording taken at 0702 from a piece of saline-soaked, air dried woven photonic cloth is shown. This reading was taken with a 2214 digital storage oscilloscope at 1x magnification and 10x amplitude. The woven photonic cloth was soaked in saline solution for three hours and dried for six hours. FIG. 5 shows two 1000 Hz frequencies, shown at A_5 and B_5, which are 8.4 Ms apart, and riding an AC interference. FIG. 6, which is a magnification of the recording shown in FIG. 5 (taken at 0710, at 10x magnification and 10x amplitude), shows the two 1000 Hz frequencies, shown at A_6 and B_6, with peak to peak separation. FIG. 7 is a magnification of the details of the first 1000 Hz frequency shown at A_6 in FIG. 6. The recording in FIG. 7 was taken at 0725 at 50x magnification and 10x amplitude. FIG. 8 is a magnification of the details of the second 1000 Hz frequency shown at B_6 in FIG. 6. The recording in FIG. 8 was taken at 0720 at 50x magnification and 10x amplification. The oscilloscope used to make the recording shown in FIGS. 5, 6, 7 and 8 was set at a 5Mv range with a 1 Ms sweep. The oscilloscope sweep shown in FIG. 8 demonstrates that the human body acts as an antenna to transmit the E. field back and forth across space as an ELF radio wave. The ELF radio signals are capable of penetrating six layers of human skin (approximately ¼" each), two feet of stacked fabric, and 2" of solid rock, with no attenuation whatsoever.

Referring now to FIG. 9, an oscilloscope sweep is shown which was taken from a six inch by fifteen inch sample of saline-soaked photonic woven cloth with one hand of a test person held approximately one foot from the photonic cloth and the thumb capacitance coupled to the 222 Tektronix digital storage oscilloscope at the cathode ray face. Two 1000 Hz frequencies are shown at C_9 and D_9 8.4 Ms apart. Both 1000 Hz frequencies have two 10,000 Hz sidebands. An example of a pair of 10,000 Hz sidebands is shown at A_9 and B_9. Similarly, FIG. 10 shows an oscilloscope sweep taken from a two inch by six inch knitted piece of saline-soaked photonic cloth with the thumb capacitance coupled to the cathode ray face and the hand of the tester directly touching the photonic cloth. The oscilloscope sweep shown in FIG. 10 demonstrates that with the body of the tester directly touching the photonic cloth, there is a tremendous increase in the amplitude of the 1000 Hz and 10,000 Hz sidebands signal, as opposed to the oscilloscope sweep shown in FIG. 9. The first main 1000 Hz signal shown at B_{10} has one sideband signal of 10,000 Hz shown at A_{10}; the second main signal shown at D_{10} also has one sideband signal of 10,000 Hz shown at C_{10}.

When the photonic cloth is placed against the human skin, the radio energy between the skin and cloth are coherent. The photonic cloth has spacial coherence because the antenna aperture is zero. The 1 Ms sweep and fixed position of the waves demonstrate that there is temporal coherence as well. Furthermore, because the signal reaches an extremely high amplitude when the photonic cloth touches the skin, the signal also becomes a phase conjugated signal.

While the invention has been particularly shown and described with reference to preferred embodiments thereof, it will be understood by those skilled in the art that the foregoing and other changes in form and details may be made therein without departing from the spirit and scope of the invention.

What is claimed is:

1. An electrode for detecting bioelectric potentials comprising a cloth of untreated natural flax and untreated natural wool and soaked in a saline solution, which, when electrically coupled to a detector, will detect radio frequencies.

2. The electrode of claim 1, wherein said cloth is woven.

3. The electrode of claim 1, wherein said cloth is knitted.

4. The electrode of claim 1, further comprising two waterproof layers encompassing said soaked cloth to maintain said saline solution within said soaked cloth.

5. The electrode of claim 1, wherein said saline solution comprises: a borate buffer system; sodium chloride; and 0.1% by weight of sorbic acid and disodium.

6. The electrode of claim 1, wherein said saline solution is seawater.

7. A photonic ionic electrode for detecting bioelectric potentials comprising a cloth of untreated natural flax combined with untreated natural wool to form the cloth, wherein the cloth is soaked in a saline solution.

8. The electrode of claim 7, wherein said untreated natural flax and said untreated natural wool are woven together.

9. The electrode of claim 7, further comprising two waterproof layers encompassing said soaked cloth to maintain said saline solution within said soaked cloth.

10. And apparatus for detecting radio frequencies, comprising: an electrode for detecting bioelectric potentials comprising a cloth constructed from untreated natural flax and untreated natural wool, wherein said cloth is soaked in saline solution;

a detector; and connection means for electrically connecting said cloth to said detector.

11. The apparatus of claim 10, wherein said cloth is woven.

12. The apparatus of claim 10, further comprising two waterproof layers encompassing said soaked cloth to maintain said saline solution within said soaked cloth.

INDEX

mantle rock, 14
manure, 29-30
    green, 29
mathematics, 55
Mayan pyramids, 85
McClintock, Barbara, 90
mechanical weathering, 23
megalithic pictograph, 72
megalithic tombs, 36, 72
    as dielectric waveguide antennae, 66
Mendel, 19
Mercer University, 80
mercuric oxide, as diamagnetic, 79
mercury, as diamagnetic, 79
mercury vapor, in fluorescent light, 55
metamorphic rock, 22
metamorphism, 22
meteorite cones, of the moon, 48
Meyer, L.D., 17
mica-minerals, 8
mica-schist, in round tower construction, 70
Middle East, 11, 18
Miles, John, 51
millet, grown by Frederick II, 18
Millikan, Robert, 90
Milton, John, 85
mineral deposits, 14
mineralization, of soil, 82
minerals, 27
    paramagnetic, 30
Mississippi, state of, 12
mitogenic rays, 89
model round towers, and insect scent
    emmissions, 87
molecules, paramagnetic behavior of, 43
Momoyama period, 47
monks, Irish, 63, 71-72
moon, as paramanetic, 48
moon rock, measurement of, 48
Mount Asama, 12
Mount Fuji, 12
Mount Hood, 11
Mount St. Helens, 11-13
mountain formation, 22
N-rays, 89
Nanzen-en stroll garden, 46
NASA, 55
Natural History, 25
Nebraska, fields of, 77
nervous system, 88
Netherlands, *x*
Nevada, and erosion, 21
"never aging rock," 39
New Hampshire, cliffs of, 35
nicotine sulphate, 18
nitrogen, as diamagnetic, 79

nitrogen fixation, 75
Nobel Prize, 89-90
noise, in radio spectrum, 64
Nordeng, Donald, 33
North American wood warblers, 9
north and south pole magnetism, 38
Northern Ireland, on the Denegal border, 5
N, P & K (nitrogen, phosphorus and
    potassium), 6
nutrients, airborne, 28
O'Brien, Dr. Edward, 80-81
oak wood, as diamagnetic, 37
oats, grown by Frederick II, 18
Ocean in the Sand, The, 33
oil drop experiment, 90
Oklahoma, fields of, 77
olives, grown by Frederick II, 18
open resonators, 65-66
Oppenheim, New York, birthplace of Mahlon
    Loomis, 52
orbital, 43
Ord, George, *vii*
Oregon, 11
ores, 14
organic compounds, as diamagnetic, 37, 46
organic molecules, as photonic oscillators, 54,
    55
Origins of Continents and Oceans, The, 15
oscilloscope, 58, 66, 68
Oxford, 58
oxygen, 28
    as paramagnetic, 29, 79-81
ozone, depletion of, 75
O'Brien, Dr. Edward, *xi*
P.C. Soil Meter (PCSM), 80-82
PACs, 75
Painter, Dr. Reginald, 77
Paiute Indians, 26
Palenque, 85
Paradise Lost, 85
paramagnetic/diamagnetic arrangement,
in geomancy, 46-47
Paramagnetic Relaxation, x
paramagnetism, *x*, 9, 27-29, 37, 43, 8
    as plant growth stimulant, 78
    definition of, 88
    in rocks, 36
    in round tower building material, 70
    in soil, 35, 79-82
    measurement of, 30
    physics of, 30
patent, first for radio, 53
"pathological science," 89
Penn State, 55
percolation, 17
Peru, 58

Also by Philip S. Callahan

My Search for Traces of God
Philip S. Callahan, Ph.D.

 Callahan's life story unfolds here along with the development of his unique blend of science, natural philosophy and spirituality. Recounting incidents from his remarkable life and career, Dr. Callahan integrates his early scientific theories and overall religious philosophy with his more recent insights regarding low-level natural energies and the nature of space and time and the realm of the miraculous. He focuses on the scientific as well as religious implications of such sacred places as Medjugorje, Lourdes and Le Puy, and discusses in detail his scientific work surrounding the miraculous images of the Shroud of Turin and the Virgin of Guadalupe. *Softcover, 202 pages. ISBN: 978-0-911311-54-9*

Tuning in to Nature
Philip S. Callahan, Ph.D.

 This 25th anniversary edition of Philip Callahan's pioneering work, updated by the author, reveals the miraculous communication systems present in nature. Learn how plants and insects communicate through emissions in the infrared frequency range and why poisonous pesticides do not solve the real problems facing agriculture. In this breakthrough book Phil Callahan uncovers why certain insects are attracted only to certain plants, the role of pheromones in nature, and how plants under stress literally signal insects to come devour them. Classic Callahan! *Softcover, 256 pages. ISBN: 978-0-911311-69-3*

Ancient Mysteries, Modern Visions
Philip S. Callahan, Ph.D.

 In this book, Phil Callahan details his initial discovery of the role and power of paramagnetic rocks in agriculture. By studying the lives, rituals and agriculture of ancient peoples, he has assembled a first-rate scientific explanation of previously misunderstood ancient practices. Learn how Egyptian priests levitated people, why rocks and soil were brought from one side of the Nile to the other, and how plants act as antennae. Fascinating reading for anyone interested in the miracles of nature and agriculture. *Softcover, 142 pages. ISBN: 978-0-911311-08-2*

To order call 1-800-355-5313
or order online at www.acresusa.com